Microbes Lethal to Mankind

by

Michael Manning

To Eugene Paul Dunne (1924 – 2014)

Even were I to rise up to Heaven, Thou art there. Even were I to descend into hell, Thou art there. Yeah, even were I to make my abode in the depths of the sea, there too would Thy right hand reach me!

An extract from: The Russian Orthodox Church Cathechism.

Published by Lulu.com

Content ID 16080313

Published by Lulu.com

ISBN: 978-1-326-13796-0

The author correctly credits all photographs used in this book. References and sources are listed in the index at the back of the book.

Front photograph a US medical doctor training in an Ebola proof hazmat suit, credit Master Sgt. Jeffrey Allen.
Back photograph Adult Deer Tick, Ixodes scapularis, Photo Scott Bauer, US Dept of Agriculture.

Michael Manning holds a B.A.Mod., M.A. and H.Dip. Education from Trinity College Dublin and is currently a freelance Science and Technical Journalist and Translator encouraging science and scientific writing in all forms of modern media.

Contents

What are Microbes?

Microbiology is the study of microorganisms - bacteria, protozoa, viruses and fungi. These organisms can only be seen under the microscope but despite their size these microbes have a massive impact on our lives. It has been estimated that there are 5 million trillion, trillion, microbial cells on Earth, which is equivalent to all the plants on the planet. Microbes constitute the largest mass of living material on earth and play a critical role in shaping the environment in which we live. Humans, plants and animals are intimately tied to the activities of microbes which recycle key nutrients and degrade organic matter. Some microbes are lethal to mankind.

Microbes guard our planet ensuring that key minerals, such as carbon and nitrogen, are constantly recycled. Even though the Earth is now populated with green plants, microbes still play a crucial role in oxygenating the atmosphere and collectively they carry out more photosynthesis than plants. Microbes degrade dead organic matter, converting the organic carbon in their bodies back into carbon dioxide.

Microbes also play a key role in the nitrogen cycle. Bacteria in the soil convert atmospheric nitrogen into nitrates in the soil. The nitrogen locked in plant and animal proteins is then degraded into nitrates by microbes and eventually converted back into nitrogen by denitrifying bacteria. Compost heaps are excellent example of how effectively microbes breakdown organic matter. Without the recycling power of microbes dead vegetation, carcasses and food waste would start piling up around us!

It wasn't until the 17th century, when the microscope was invented by Robert Hooke, that the existence of microbes was even suspected. Hooke's microscope, however, could only achieve magnifications of 20-30 times - not powerful enough to see bacteria. Around 1668 Antoine van Leeuwenhoek, an amateur microscope builder, improved microscope design so that he was able to make a microscope capable of magnifications of up to 200 times. Van Leeuwenhoek examined pond water, tooth scrapings and then almost anything else he could investigate. In 1683 he described, in a letter to the Royal Society, how he saw 'an unbelievably great company of living animalcules, swimming more nimbly that I had ever seen up to this time' he looked at the tooth scraping from an elderly man, who had never cleaned his teeth - the animacules were bacteria.

Microbiologists discovered that microbes are found everywhere. Microbes are an incredibly diverse group of organisms and can grow in extreme environments that no other living organisms can tolerate. Bacteria have been found to thrive in volcanic hot springs, where temperatures typically reach near boiling point. At the other extreme, living bacteria have also been discovered in Antarctic deserts, where temperatures range from -15 to -30°C. Bacteria can also thrive in salt flats, pools of saturated brine, where salt concentrations range from 120 to 230 grams per litre. Bacteria living in these inhospitable environments are described as extremophiles.

Bacteria growing in extreme conditions have proved to be a rich source of enzymes for the biotechnology industry. Fat-degrading and protein-degrading enzymes from bacteria isolated from hot springs have been used to make biological washing powders. Clearing up oil spills in cold oceanic environments, the production of ice cream and artificial snow have also benefited from enzymes, produced by bacteria thriving in near zero temperatures.

Microorganisms play an important role in producing a whole variety of delicious foods. Microbes are used to in make chocolate. Cocoa pods are split open and their contents – 20 to 30 bitter seeds in a sweet sticky pulp - are heaped together and covered with banana skins and naturally fermented for 7 days. Over 30 different types of bacteria are involved in this process, along with yeasts and moulds. The sticky pulp becomes a turbid chocolate-coloured broth which gives the cacao seeds both their characteristic chocolate flavour and colour. Microbes play a key role in making wine and beer and foods such as bread, cheese and yoghurt. Salt-loving bacteria, like those found in salt flats, play a key role in the production of Japanese soy sauce. Chinese cooking depends on microbes which are essential in the production of black bean and yellow bean sauces. Salt-loving bacteria are involved in the production of cured meats and sausages such as salami. Without microbes our culinary repertoire would be smaller and our diets extremely bland.

The interaction between microbes and food is not always beneficial. Mouldy bread and rotten fruit caused by microbial degradation is alarming. Pizzas past their sell by date may not be obviously full of bacteria, but sell by dates are based on the amount of time it takes for the numbers of bacteria to reach a level where the chances of food poisoning are high. Microbial spoilage makes food unappetising and perhaps foul tasting, but rotten food won't automatically make you sick – but contamination of food with microbial pathogens will.

The bacterium *Campylobacter jejunii* is the commonest cause of food poisoning in Europe. In 2010 at least 65% of all fresh chickens sold in the Europe were contaminated with *C. jejunii*. Cooking kills this bacterium yet it is still responsible for many cases of food poisoning annually. For most healthy people, food poisoning is not life-threatening. For both babies and elderly people food poisoning can be extremely serious as it can cause severe dehydration and kidney failure. Learning how to handle raw meat and how to cook poultry and meat properly is therefore essential, particularly during the barbecue season when the incidence of food poisoning, due to poorly cooked food, soars.

At the beginning of the 20th century infectious diseases, caused by microbial pathogens, were the major cause of death. Large numbers of children and the elderly succumbed to diseases such as tuberculosis, diphtheria and pneumonia. At this time microbiologists had little idea about how diseases were spread, or how they could be controlled, so epidemics flourished. The Spanish influenza pandemic of 1918–1919, caused c.50 million deaths worldwide - more than the total number of deaths recorded in World War One. Diarrhoeal disease was also common since people regularly ate contaminated food and drank contaminated water.

Microbiology research has been concerned with developing antibiotics and vaccines to protect the population from infectious disease. The discovery of penicillin by Alexander Fleming saved the lives of many millions of people. The development of vaccines which protect against, diphtheria and pneumonia dramatically reduced the number of childhood deaths caused by these diseases. Children in developed countries are also routinely vaccinated against common viral infections such as measles, mumps, rubella and polio. As a direct result of the efforts of microbiologists, smallpox, once a dreadful scourge, is now officially extinct on the planet. However a vaccine against HIV, which in 2009 was reported to infect approximately 33.3 million people around the world, still eludes us.

Babies are colonised by bacteria immediately after birth. It has been estimated that the average person is colonised by 200 trillion bacteria, comprising at least 1,000 different species. The bacteria that call the human body home are often essential for our health and well being. Our intestines contain about 100 trillion bacteria and collectively they make up 60% of the dry weight of faeces. These intestinal bacteria play an essential role in helping us to digest food, they provide us with essential vitamins such as vitamin K and biotin and they help to prevent the growth of harmful pathogenic bacteria. The surface of our skin is also home to millions of friendly bacteria which crowd out potential pathogens and prevent them from growing. One bacterium which is abundant on the skin is *Staphylococcus epidermidis* which produces chemicals called bacteriocins that kill pathogenic bacteria. Friendly bacteria can also be found in our noses but many of these bacteria also carry a health warning. *Neisseria meningnitidis* which causes meningitis, lives in the noses of millions of people without causing disease, but if the immune system becomes weakened through ill-health then this bacteria can, almost by accident, cause disease which may result in the death of the human that has become its home.

The articles in this book explain how microbes adversely affect mankind. A number of cancers are discussed due to their high mortality rate.

Sulphur Bacteria in Mid-Ocean Hot Springs and Possible Life on Saturn's Moons

Tube worms in deep sea vents enjoy a long life in ocean floor smoker vents. Image courtesy of University of Washington; NOAA/OAR/OER

The odd animals inhabiting hot springs in the deep oceans, and subglacial ecosystems in polar regions, may have been living there since at least the Late Cretaceous Period – simply because of their strangeness.

Almost every one of the animals discovered in these extreme environments were new to science when found, and cannot live elsewhere.

The vent animals, for example, are not just different species from other deep sea animals; scientists actually needed place them into new phyla simply to accommodate their strangeness.

Could extremophile bacteria indicate the possibility of similarly alien life forms in the unforgiving and extreme climates of other planets, or their moons?

The Chemolithotrophic Bacteria in Galapagos Hot Springs Tubeworms

In 1977, when divers in the submersible *Alvin* found the waters of the Galapagos hot springs teeming with sulphur-eating bacteria, they quickly concluded that chemosynthetic bacteria must be the nutrient that supported this new ecosystem.

Chemosynthesis is the process by which organisms use inorganic molecules as a source of energy, rather than the sunlight used in photosynthesis.When scientists examined the giant tubeworms *Riftia pachyptila,* they found neither a mouth nor a gut – this meant that the giant tubeworms were an entirely new kind of organism.An entire community of animals in symbiosis with chemolithotrophic bacteria, feeding by means of geochemistry, was an exciting new find, and revealed that extremophile organisms could survive in environments lethal to most living organisms.

Polar Extremophiles

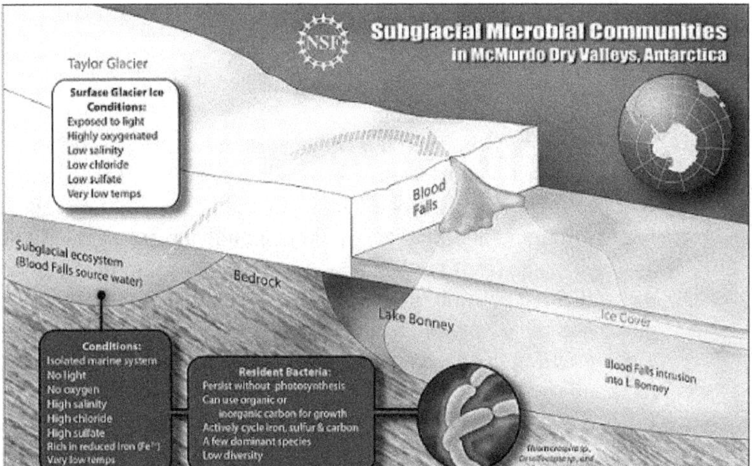

Subglacial microbial communities have survived in cold, darkness and absence of oxygen for a million years in McMurdo Dry Valleys, Antarctica: Image by Zina Deretsky, US NSF

Freezing-cold polar environments on Earth also harbour extremophile bacteria; extremophiles are found in glaciers and on high altitude peaks, fully functioning in these extreme environments.

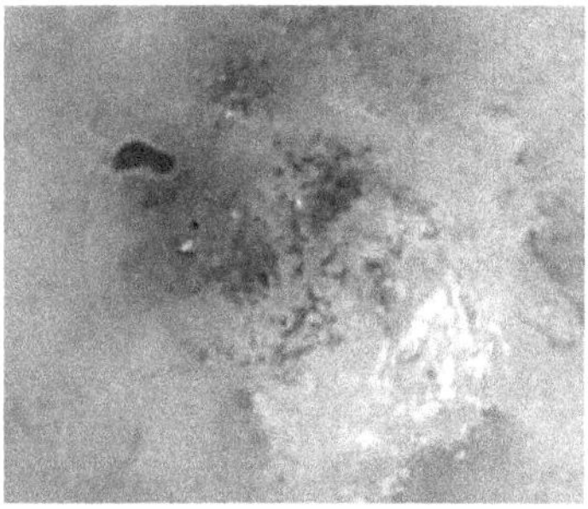

Titan's South Pole clearly shows a methane lake. NASA/JPL Space Science Institute

Life on Saturn's Moons: Enceladus, Titan, and Jupiter's Europa?

Understanding the oddities of life in our most extreme regions leads us to wonder: is life on Enceladus, Titan, or Europa possible?

- Enceladus is the brightest moon in our solar system. It is composed entirely of ice, fully reflecting light. In 2005, the Cassini spacecraft photographed geysers of ice and water vapour being expelled at least 300km into space, so there must be liquid water under the moon's icy surface, and sufficient heat internally to propel ice into water vapour.
- Titan is the only moon in our Solar System that has a substantial atmosphere (it is mostly nitrogen). When the Cassini spacecraft dropped the Huygens probe into Titan's atmosphere, the probe found ammonia and methane. Lakes of methane are clearly visible, as you can see from the image above. Ammonia and methane could theoretically combine, in an electrically charged environment, to make organic compounds, and extremophile bacteria, similar to those found in deep ocean hot springs, could survive in this methanological system.
- Europa, the second moon of Jupiter, has an icy surface with a suspected liquid ocean below the surface. Life similar to that of subglacial regions is possible on Europa as well.

Extremophiles Provide Insight into Potential Off-Planet Life Forms

Deep ocean hot springs and sub-glacial regions harbour extremophile bacteria, and Enceladus, with abundant water vapour geysers; Titan, with a methane and nitrogen atmosphere; and Europa, with its icy surface, could theoretically harbour extremophiles as well.

The potential for life occurring twice in our Solar System is exciting; Cassini's ongoing mission to photograph Titan's surface may provide more insight over time.

References

NASA. *Cassini Solstice Mission*. (2012). Accessed July 22, 2012.

Museum of Science. *Extremophile Bacteria in the Arctic Circle and on Titan*. (2010). Accessed July 22, 2012.

Kunzig, R. *Mapping the Deep, The Extraordinary Story of Ocean Science*. (2000). W. W. Norton & Company/Sort of Books.

Cavanaugh, Colleen et al. *Prokaryotic Cells in the Hydrothermal Vent Tube Worm Riftia pachyptilaJones: Possible Chemoautotrophic Symbionts*. (1981). Science. 213 (1981): 340-342. Accessed July 22, 2012.

New Flu Treatment May Prevent Pandemic Influenza Outbreaks

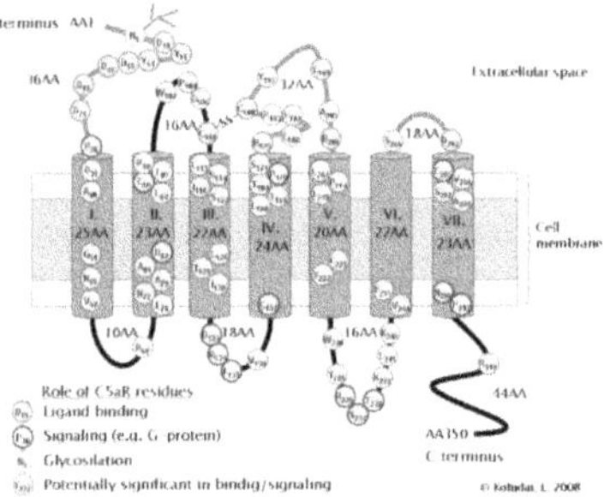

Preventing the flu via the C5a receptor: Image by Kohidai, L.

Are pandemic flu fears a thing of the past?

Flu research published today in the Public Library of Science demonstrates that a new chemical created to bind to the receptors of an immune protein called C5a, or Complement Component 5a, may be the answer to our influenza worries.

C5a plays an essential role in airways defence against bacterial, viral, and fungal infection.

Now, researchers from the University of Nebraska Medical Centre and San Diego have developed an agonist, or chemical that binds to the C5a's receptor, and retains the cell's immune-enhancing activity, while eliminating its inflammatory properties. In this research, mice treated with the chemical, called EP67, within twenty-four hours of infection with the flue were significantly protected from influenza-induced weight loss.

EP67 delivered twenty-four hours after lethal infection completely blocked influenza-induced mortality; this chemical essentially prevented death from the flu.

EP67 may prove effective against a wide variety of viral, bacterial, and fungal pathogens; potentially stopping the worldwide spread of respiratory diseases, including novel strains of influenza.

Flu Research: Purpose of the Test, and Methodology

The development of therapeutic agents to combat infection by enhancing the immune system could result in protection against multiple pathogens, instead of the current "one bug, one drug" system in which each individual strain of a virus requires its own unique vaccine or treatment.

In this research, mice were infected with influenza virus strain H1N1. The virus count was calculated using uninfected lung tissue from the mice, to which known numbers of influenza particles called virions (virions are made up of an outer protein shell and an inner core of nucleic acid from the virus) had been added prior to its isolation.

Influenza Virus Research: Results

The results of this research were significant. Key findings included:

- EP67 induces an immune response in the airway: This chemical induced an immune response two hours and two days after the mice were treated with EP67.
- Significant protection from flu-induced weight loss: Mice treated with EP67 within a day of infection were significantly protected from influenza-induced weight loss on day eight. The immune response induced by EP67 in the presence of influenza infection was similar to that seen in mice treated with EP67 in the absence of an influenza infection.
- Protection from flu death: Delayed EP67 treatment protects from lethal influenza infection.

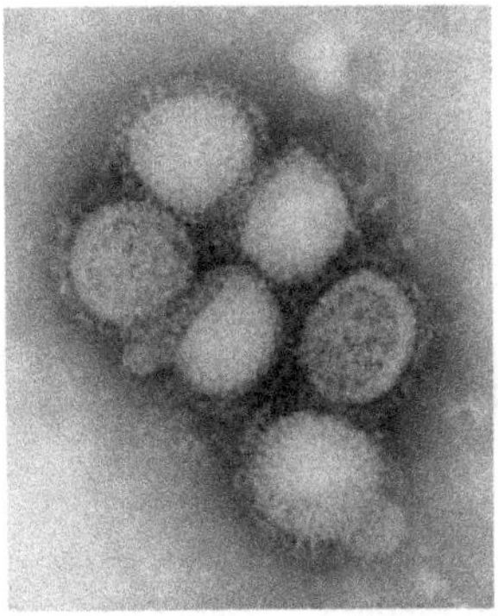

The potential pandemic influenza virus may be a variant of the H1N1, or Swine Flu virus. Photo Credit: C. S. Goldsmith and A. Balish, CDC

Infection Treatment: Summary

The ability to enhance the immune response is the prime objective when treating viral, bacterial, and fungal infections. This strategy requires development of immune proteins which induce innate immunity, without causing inflammation, or other side effects. C5a is integral to the immune response, as it attracts inflammatory cells to the site of infection, and activates their immune responses. This study shows that the delivery of EP67 to the airways induces a potent anti-viral response, characterized by an influx of immune response cell populations. A single dose of EP67 ensures that a patient's innate immune response will provide 100% protection following infection with what would normally be a lethal dose of influenza.

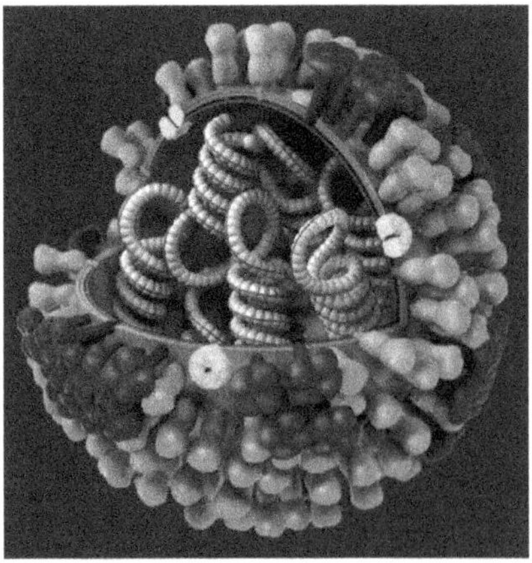

Could EP67 wipe out fears of the Influenza Virus? Image courtesy of the US CDC

Protection from the Flu: Significance of Influenza Research

According to Dr. Phillips, "*The potential to induce immediate immunological protection, even prior to identifying a specific pathogen, is pretty significant. This has implications for global health and bioterrorism, and even the world food supply. EP67 works in mammals and birds, so it's therapeutic potential extends beyond human respiratory disease into veterinary medicine. Work focused on bioterrorism talks about human pathogens, but protecting the world food supply is an equally important concern.*"

Pandemic Flu Fears a Thing of the Past?

The results indicate that EP67 may function as a broad-spectrum emergency therapeutic for diverse respiratory infection.

Decoded Science asked Dr. Joy Phillips, co-author of this study, whether EP67 could be made available to counter SARS, or an influenza pandemic. Her response was an unequivocal, "*YES!*"

Dr. Phillips goes on to tell us, "*I absolutely think that EP67 can be used as an emergency treatment against influenza or SARS. In fact, I think it has huge potential as an emergency therapeutic following exposure to a myriad of respiratory infections, even prior to their identification. This could be a pathogen exposure, a bioterrorism setting, or with the advent of new global pathogens.*"

EP67: Protection from the Flu This research was designed to determine whether EP67, previously used primarily as an immune-enhancer, could provide direct protection in the face of respiratory infection. The results indicate that the stimulation of cells in the airways provides immediate protection, even from an established infection, without compromising the acquired response to infection. These findings may revolutionise flu treatments as well as remove fears of Pandemic Flu outbreaks such as Swine Flu and Bird Flu.

Resources:Sanderson, S., et al. *Innate Immune Induction and Influenza Protection Elicited by a Response-Selective Agonist of Human C5a*. (2012). PLOS One. Accessed July 6, 2012.

Ebola Virus Outbreak in Uganda July 2012 is a Cause for Concern

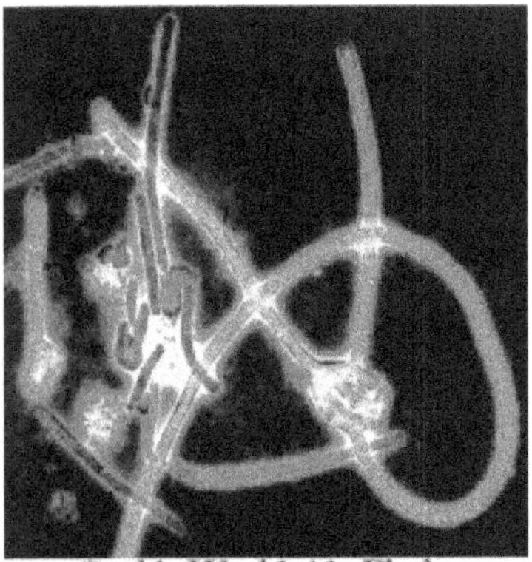

Could a Worldwide Ebola Outbreak Start in Uganda?

Ebola Virus particles which cause Hemorrhagic Fever. Image by Thomas W. Geisbert, Boston University School of Medicine

The deadly Ebola Hemorrhagic Fever (Ebola HF) virus claimed a total of 16 lives since the outbreak began in western Uganda in the first week of July 2012, according to World Health Organization reports, but there were no incidents of the infection spreading in the capital, Kampala.

Ebola, one of the world's most virulent diseases, is spread through close personal contact. Uganda's President Museveni has said that health officials are trying to trace people who were in contact with victims so that they could be quarantined – but is the potential for a worldwide Ebola outbreak, as dramatised in the 1988 Outbreak film with Dustin Hoffman, really possible?

About the Ebola Virus

Ebola HF is a fatal disease in humans and primates that has appeared intermittently since its initial recognition in 1976.

The disease is caused by Ebola virus, named after a river in the Democratic Republic of the Congo in Africa, where it was first recognized in 1976.

The virus is an RNA virus classified as the Filoviridae.

There are five identified subtypes of Ebola virus, and four of the five cause disease in humans: Ebola-Zaire, Ebola-Sudan, Ebola-Ivory Coast and Ebola-Bundibugyo. The fifth, Ebola-Reston affects nonhuman primates.

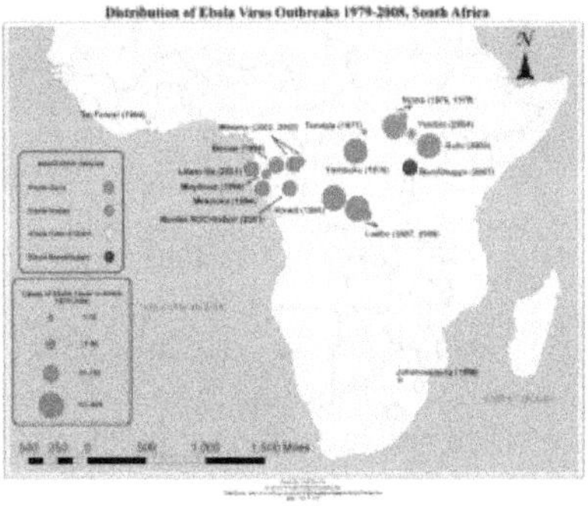

Ebola virus incidents in Central Africa 1979-2008 Image by Zach Orecchio

Where has Ebola Been Identified?

Confirmed cases of Ebola HF have been reported in the Democratic Republic of the Congo, Gabon, Sudan, the Ivory Coast, Uganda, and the Republic of the Congo. Check the map for specific locations and additional details regarding outbreaks from 1979 through 2008.

How Does the Ebola Infection Spread?

Ebola HF appears in sporadic outbreaks, often in a health-care setting. Isolated cases do occur, but are often unrecognised. In the case of the current outbreak, the symptoms of the disease were not immediately recognized as Ebola, which may have increased the spread potential of the outbreak. Infections with Ebola virus are acute, and lack a carrier state. The natural reservoir of the virus is unknown, however researchers believe the initial infection is through contact with an infected animal.

After the initial infection, human victims are exposed to Ebola virus through direct contact with the blood of an infected person. The virus is spread through families and friends, or other situations in which people are in close contact with the secretions of an infected person and contaminated objects such as needles.

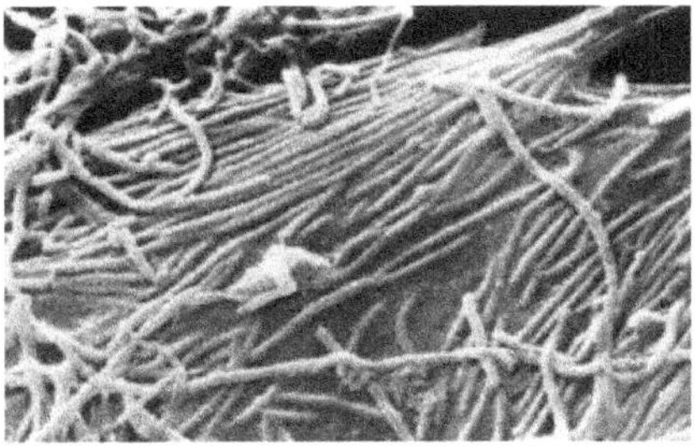

Ebola spreads through close contact. Image courtesy of "Charting the Path of the Deadly Ebola Virus in Central Africa." Published in PLoS Biol 3/11/2005: e403

Ebola HF transmission may occur in African clinics, or hospitals, where patients are treated without adequate protection. When needles or syringes are used, they may not be sterilised and are often reused due to shortages, so, infection can spread rapidly.

What Are The Symptoms of Ebola Hemorrhagic Fever?

The incubation period for Ebola HF ranges from 2 to 21 days. The onset of illness is characterized by fever, headache, joint and muscle aches, sore throat, and weakness. These symptoms are followed by diarrhoea, vomiting, and stomach pain. A rash, red eyes, hiccups, and internal and external bleeding may be seen in some patients. Some people recover from Ebola HF – patients who die have not developed an adequate immune response to the virus.

Uganda Today

Although other nations continue to provide aid to Uganda, the Irish government ended their annual humanitarian aid programme in 2010, due primarily to the fact that the aid was not reaching the people in most need. A European health worker, (who wishes to remain anonymous), worked near the capital Kampala during July 2012, and described widespread corruption in a private interview with Decoded Science. When it comes to serving the people of Uganda, workers report challenging circumstances, including men with multiple wives and no fatherhood responsibility, and high rates of HIV.

Could The Uganda Ebola Virus Spread Worldwide?

The international airport in Kigali, Rwanda, which provides international air access for East Africa, could, in theory, carry an individual infected with Ebola HF worldwide. Although health officials are working to isolate all potentially-infected individuals, this infection does have the potential to harm the more than 7 billion humans living on our fragile planet.

Luckily, this outbreak of Ebola HF appears to be self limiting – unlike fictional accounts, such as the 1988 Hollywood film *Outbreak* with Dustin Hoffman, which dramatised the worldwide threat of an Ebola HF outbreak.

References

USAID. *Uganda*. (2012). Accessed August 10, 2012.

Muhumuza, R. *Ebola Under Control in Uganda*. (2012) Boston.com. Accessed August 10, 2012.

US CDC. *Ebola Hemorrhagic Fever*. (2012). Accessed August 10, 2012.

World Health Organization. *Ebola in Uganda: August 10 Update*. (2012). Accessed August 10, 2012.

BBC News. *Outbreak of Ebola in Uganda*. (2012). Accessed August 10, 2012.

A New Approach Towards Antibiotic Resistant Tuberculosis

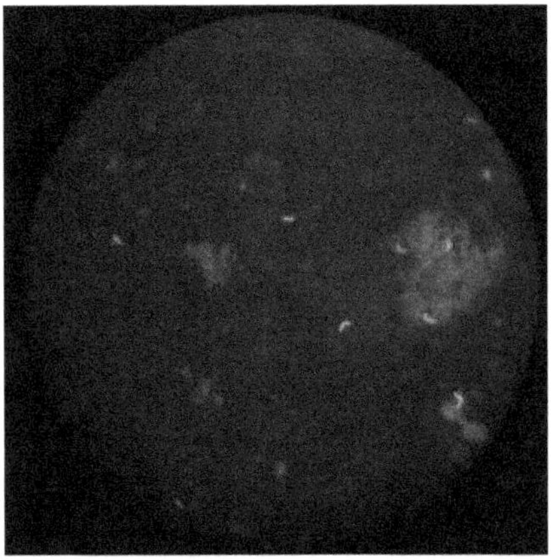

Result of a FDA-test under the microscope: the fluorescent lines are living tuberculosis bacilli, on a background of cellular debris from human sputum. Image © ITG, reproduced with permission.

Tuberculosis (TB) is a bacterial infection caused by the organism, *Mycobacterium tuberculosis,* and it's deadly.

Not only that, due to the problems with determining the appropriate antibiotic, multi-resistant tuberculosis has the potential to become a worldwide epidemic, negating all medical achievements of the last several decades.

The good news: a new approach to finding antibiotic-resistant TB could dramatically improve treatment options.

TB Facts

In humans, the bacillus (rod-shaped bacterium) thrives in environments where the oxygen tension is relatively high, such as the rounded upper part of the lungs, the renal parenchyma, and the growing ends of bones.

Infection may lead to lung necrosis and cavitation. A person with untreated pulmonary tuberculosis is infectious, and may transmit the organism to those with whom he is in close contact.

Tuberculosis: Scourge of the 19th and 20th Century

A century ago, tuberculosis was more terrifying than cancer is today. Over the nineteenth and twentieth centuries, a billion people died from TB – more than the world population in 1800. During the 1950s, the disease was interdicted using newly developed antibiotics. European sanatoria were closed and converted into hotels. Today the industrialized world does not comprehend the gruesome nature of consumptive disease. The treatment was so successful that the World Health Organization (WHO) in 1960 decided to eradicate tuberculosis once and for all. It almost worked.

Persistent Resistant Tuberculosis

Mycobacterium tuberculosis is persistent, demanding treatment with several antibiotics simultaneously for months, which is impossible in developing countries. Erratic or halted treatments led to growing numbers of bacilli which were resistant to several antibiotics. In the early 1980s, the death toll stagnated, then, rose inexorably. The arrival of AIDS during the 1980s caused grave concern, because infection with AIDS made the patient more susceptible to TB.

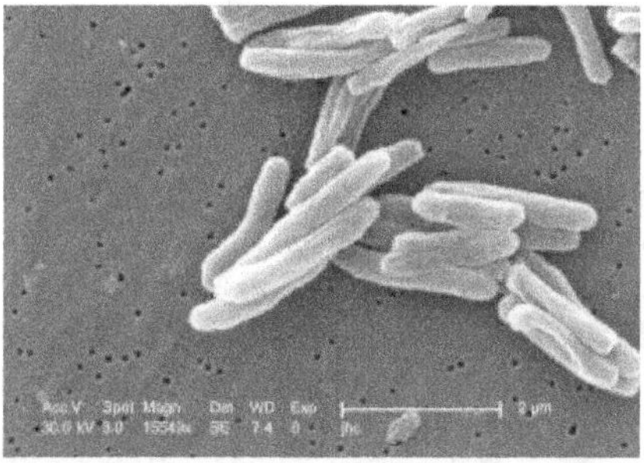

Mycobacterium tuberculosis is a deadly disease, and can be very resistant to antibiotics as well. Image courtesy of the US CDC

TB Treatment in 2012

Today, there is a growing incidence of multi-resistant tuberculosis, withstanding the best antibiotics, which is only treatable with costly toxic drugs, which patients in developing countries cannot afford.

According to WHO, of the 5 million or so multi-resistant cases during the last decade, only one percent had access to treatment. In 1991, for example, a tuberculosis outbreak in New York was resistant to 11 antibiotics, and cases have also been reported where every antibiotic was ineffective.

Thankfully, these omni-resistant bacilli perished with their hosts before they could spread.

In 2012, 1 in 30 of new TB cases worldwide were multi-resistant, with some incidents of 1 in 3. Patients relapsed after a first treatment, with on average, 1 in 5 being multi-resistant, with peaks up to 65%. The highest numbers were registered in the former Soviet Union.

Resistant cases, which do not react to normal treatments, need to be recognised as early as possible, and immediately treated with effective second-line antibiotics. The laboratory tests to identify resistant TB bugs are cumbersome – the WHO estimates that in 2009 only 11% of multi-resistant cases were actually discovered.

Tuberculosis bacilli resistant to major antibiotics are a serious threat to world health, but now, scientists of the Antwerp Institute of Tropical Medicine have redeveloped a forgotten technique which detects resistant tuberculosis in circumstances where this was not previously feasible.

The Test for Multi-resistant TB

Checking smears under the microscope remains the recommended technique for TB screening, but it cannot differentiate between living and dead bacilli, so the bacilli found may be the cadavers of a successful treatment, or resistant survivors. If the numbers don't fall after multiple tests, then the bacilli is identified as a resistant strain, with the patient remaining contagious.

High-tech PCR technology immediately ascertains whether the bacillus is from a resistant strain, but in practice and certainly in resource-limited countries this is unfeasible. It also is impossible to cultivate every sample and then bombard it with every possible antibiotic to survey which ones still work for that individual patient.

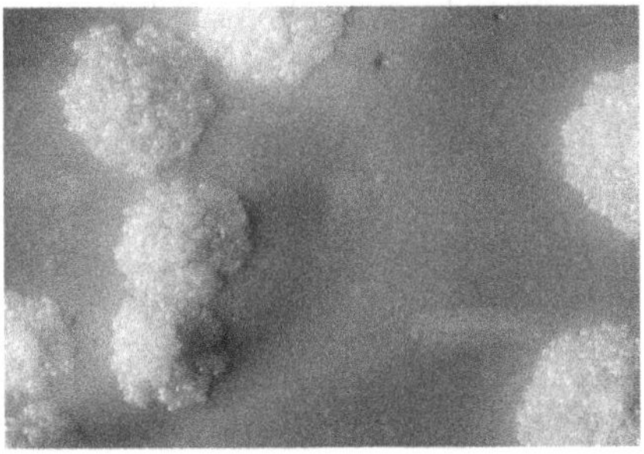

A TB culture shows the colonial nature of the bacterium. Image courtesy of the US CDC

Dr. Armand Van Deun and colleagues therefore gave a new application to a forgotten technique: vital staining with fluorescein diacetate (FDA).

It only stains living TB bacilli, so doctors can immediately see those bacilli that are escaping treatment.

The scientists improved the detection of the luminous bacilli by replacing the classical fluorescence microscope with its LED counterpart.

Together with colleagues in Bangladesh, they tested the approach in the field for four years in a study made possible by a grant from the Damien Foundation.

Their approach appears to be much more efficient for poor countries. If, after treatment, the FDA-test were negative, in 95% of cases more elaborate tests didn't find active bacilli in the patient's sputum either.

If the test were positive, the patient was identified as the carrier of a resistant bacillus.

A Simple, Less Costly, Fluorescein Test Allows for Correct Second Line Treatment of TB

This simple test allows the detection of a high number of resistant TB bacilli that otherwise would not have been discovered. The scientists report in the *International Journal of Tuberculosis and Lung Disease* that three times more patients could directly switch to the correct second-line treatment without losing time on a regimen ineffective against their resistant bacilli. This rediscovered technique can cut in half the number of cases where doctors start a retreatment, because it ascertains that the bacilli detected by the classical microscopy in fact are dead ones, which do not require further treatment.

The result? A more efficient treatment system, fewer deaths, and less spread of the disease. If this test is used consistently, it could be the turning point in the fight against TB around the world.

References

Van Deun, A., Maug, A. K. J., Hossain, A., Gumusboga, M., de Jong, B. C. *Fluorescein diacetate vital staining allows earlier diagnosis of rifampicin-resistant tuberculosis*. (2012). International Journal of Tuberculosis and Lung Disease. Accessed September 5, 2012.

United States Center for Disease Control and Prevention. *Report on Tuberculosis*. (2012). Accessed September 5, 2012.

World Health Organization. *Report on Tuberculosis*. (2012). Accessed September 5, 2012.

Rodent Borne Hantavirus Pulmonary Syndrome (HPS) in Yosemite National Park

Deaths from Hantavirus Infections in the United States since 1993 – Image courtesy of the US CDC

A recent outbreak of hantavirus infections among campers in the Yosemite area has left three dead out of a total of eight cases, as health officials work to contact others who may have been exposed to the disease.

What is Hantavirus anyway, and how does it affect the human body?

Hantavirus, and Hantavirus Pulmonary Syndrome

Hantavirus is carried in rodent faeces, urine and saliva which dries out and mixes with dust, and is inhaled by humans, particularly in small, confined spaces with poor ventilation.

Hantavirus infections may develop into Hantavirus Pulmonary Syndrome (HPS), which can be fatal.

Hantaviruses are negative RNA viruses from the Bunyviridae family, and can infect people through contact with hantavirus infected rodents, or their urine and droppings.

The Sin Nombre hantavirus was first recognized in 1993, and is one of several New World hantaviruses circulating in the US.

Yosemite Hantavirus Outbreak

The recent deaths of three people in Yosemite National Park, where at least 10,000 people stayed in rodent-infested huts during the summer of 2012, may be self limiting. Most of the victims are believed to have contracted the virus while staying in tent-style cabins during the summer of 2012 in a popular camping area called Curry Village. Park officials closed 91 tent-cabins on finding deer mice which burrow through holes to nest in the double walls of the cabins.

Park authorities notified 22,000 visitors to Yosemite, who rented the tent cabins from June through August, that they may have been exposed to hantavirus, and experts continue to investigate the outbreak. The number of cases could rise, as visitors exposed to the virus become ill over time.

Hantavirus and Medical Care

Hantavirus symptoms can be severe or mild, but there is a standard battery of tests performed by medical personnel when hatavirus infection is suspected.

- Symptoms: The early symptoms of hantavirus disease are similar to the flu: chills, fever and muscle aches – although people with hantavirus may improve for a short time, within 1-2 days breathing becomes laboured, and the disease takes hold rapidly. Later, more severe, symptoms include: dry cough, general ill feeling, headache, nausea and vomiting, and shortness of breath.

- Examinations and Tests: When a doctor examines the patient, he or she may find acute respiratory distress syndrome (ARDS), kidney failure, low blood pressure, and low blood oxygen levels causing the skin to turn blue (cyanosis). The doctor will most likely order the following tests: blood tests to confirm hantavirus, full blood count, complete metabolic indicators, kidney and liver function to confirm symptoms.

- 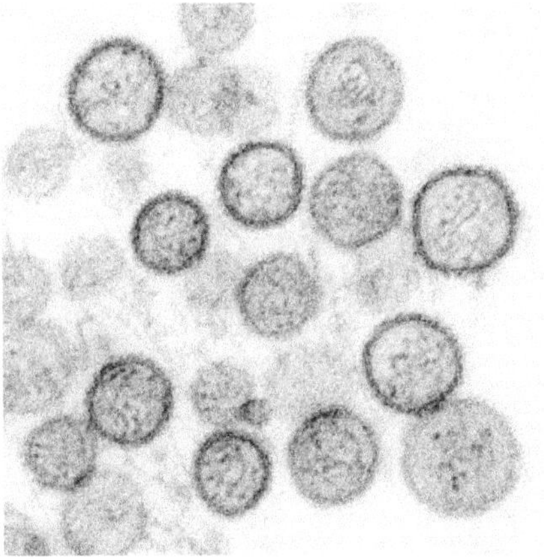

Transmission Electron Micrograph Sin Nombre Hantavirus – Image courtesy Cynthisa Goldsmith and Luanne Elliott, CDC US Govt

Hantavirus Treatment and Prognosis

People with hantavirus are admitted to the intensive care unit (ICU) where treatments include: oxygen, breathing tube or breathing machine in severe cases, and administration of Ribavirin to treat kidney-related problems and reduce the risk of death.

There is no effective treatment for hantavirus infection in the lungs.

Hantavirus is a serious infection that can rapidly get worse, resulting in lung failure and death. Possible complications include kidney failure, heart failure, and lung failure. Despite aggressive treatment, at least half of people with hantavirus in their lungs die.

Current Hantavirus Outbreak Deaths

At least eight people have been infected with Hantavirus Pulmonary Syndrome, and three people have died. As the United States CDC has assured the public, this disease has a self-limiting nature, since no cases of transmission between humans have ever been found.

References

National Institute of Health. National Institute of Health. (2012). Medlineplus. Accessed on September 12, 2012.

NBC News Services. 10,000 at risk of hantavirus in Yosemite Outbreak. (2012). Accessed on September 12, 2012.

U.S. Centers for Disease Control and Prevention. August 2012 – Yosemite National Park Outbreak 2012. Accessed on September 12, 2012

Hospital Acquired MRSA Infections: Hazardous to Hospital Inpatients.

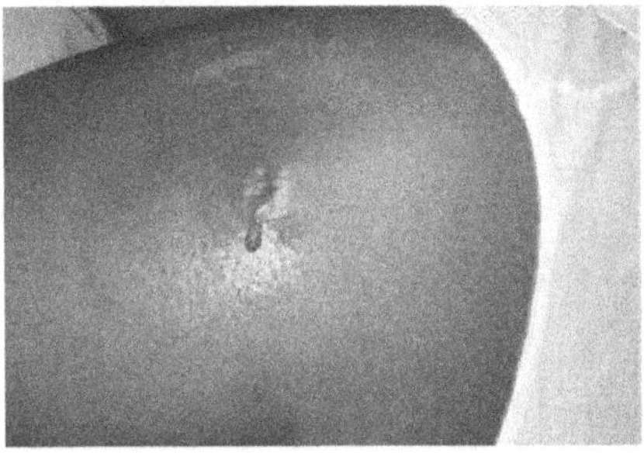

Cutaneous abscess MRSA Staphylococcus aureus in the hip of a prisoner 2005, Courtesy US CDC

Hospital Acquired-Methicillin Resistant *Staphylococcus aureus* (HA-MRSA) infections, resistant to all penicillin-type antibiotics, are a major problem in today's hospitals.

Researchers are working to resolve this life-threatening issue by studying effective hygiene methods in an effort to combat these infections.

Interestingly, a recent study on this topic found instead that HA-MRSA infections decreased when antibiotic usage in the hospital was reduced.

MRSA and Antimicrobial Resistance

MRSA infections in hospitals may prove lethal to patients recovering from life-saving operations; those on immunosuppressive medication, and even in maternity wards.

HA-MRSA infections do not respond to the normal range of antibiotics, and patients often succumb to simple cutaneous abscesses such as the one pictured, which may develop into severe systemic infections. Resistant micro-organisms persist in hospital settings.

The Ten Year, MRSA Study at St. George's Hospital, London, U.K.

Medical researchers at St. George's Hospital, University of London, U.K., monitored MRSA infection over 10 years for a study published in late September, 2012. The study found a reduction in MRSA rates – but it wasn't the result of improved hygiene; the reduction coincided with a drop in hospital prescriptions of ciprofloxacin, a member of the fluoroquinolone family of antibiotics. Main points from the study are as follows:

- Successful HA-MRSA clones such as CC22 SCC*mec*IV are resistant to both fluoroquinolones and many additional antibiotics.
- The prescribing of antibiotics in hospitals clearly contributes to the spread of HA-MRSA infections.
- Improved hygiene and hand washing had little effect on reducing MRSA infection rates during the study period.
- Significantly, when ciprofloxacin prescriptions were reduced by 70% during the same study period, the incidence of MRSA fell by 50%, and remained at this level.

How were MRSA Infections Reduced?

According to author Dr. Jodi Lindsay, at St George's, University of London, "*the study results suggest HA-MRSA infections rely on ciprofloxacin, and fluoroquinolones in general, to thrive in hospitals.*" Dr. Lindsay's interpretation of the findings is that controlling MRSA superbugs requires finding alternative ways to use antibiotics, rather than simply focusing on infection control techniques.

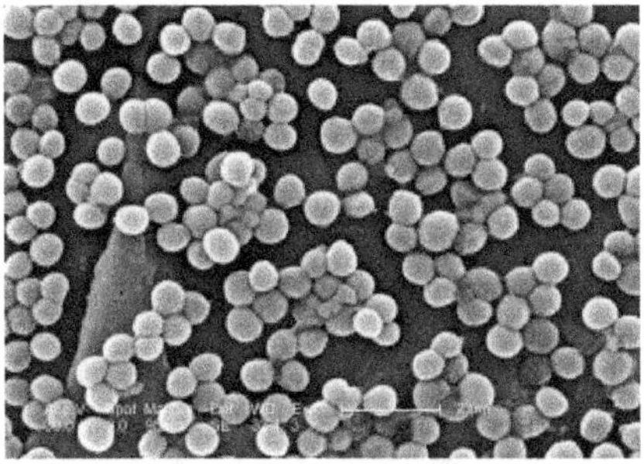

Will hospitals be able to reduce MRSA infections by lowering antibiotic prescriptions? Image by Janice Haney Carr, Centers for Disease Control and Prevention

C22 MRSA- the Superbug

The researchers identified C22 as the dominant strain of MRSA which develops and maintains multi-drug resistance and which adapts to persist on hospital surfaces.

According to author Dr. Lindsay, "*studying the dynamic of how MRSA bacteria strains evolve in hospitals in response to infection control and antibiotic prescribing, is essential to the process of determining which strains are likely to have the best long-term outcomes.*"

The bacteriophage is a potential anti-MRSA measure.

A bacteriophage is a virus that kills bacteria and could counter intractable MRSA infections in patients.

Dr. Lindsay told Decoded Science that, *"Bacteriophage has potential to combat MRSA, but still needs substantial development before a product is likely to reach the market."*

MRSA Study Highlighted Reduced Infection Rates

This study examined whether improved infection control measures contributed to this decrease in MRSA infection. However, during a four-year period, the only major reduction in MRSA infection rates coincided with the reduction in ciprofloxacin prescriptions. In fact, the only measure that appreciably reduced deadly MRSA infection was the reduction of prescription antibiotics.

References:

World Health Organization. *MRSA: Antimicrobial Resistance*. (2012). Accessed October 21, 2012.

Gwenan M. Knight, G., Budd, E., Whitney, L., Thornley, A., Al-Ghusein, H., Planche, T., Lindsay, J. *Shift in dominant hospital-associated methicillin-resistant Staphylococcus aureus (HA-MRSA) clones over time*. (2012). Journal of Antimicrobial Chemotherapy. Accessed October 21, 2012.

United States CDC. MRSA Statistics 2010. (2011). Accessed October 21, 2012.

Irish Health Service Executive. IrishHealth,22 Babies infected with MRSA. (2012). Accessed October 21,2012.

Hodgkin's Lymphoma: New Treatments for Cancer of the Lymph Nodes

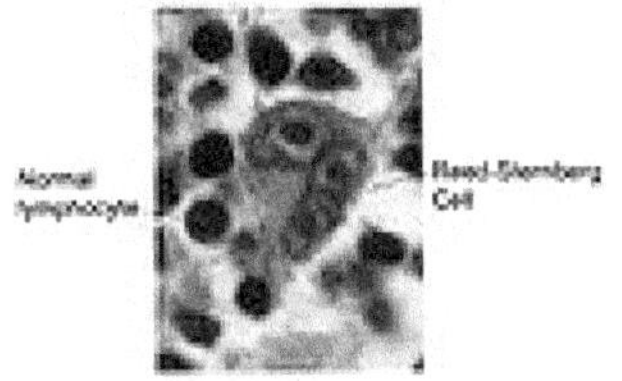

Micrograph of a lymph node with Hodgkin's lymphoma. Image by Nephron

One in five of all lymphomas, a cancer that starts in the immune cells of the lymph nodes, are Hodgkin's lymphoma.

This cancer primarily affects people who are between 15 and 35 years old, or over 55 years old.

The disease begins when B-lymphocytes, the immune cells that produce antibodies, become abnormal, creating Reed-Stemberg cells.

These cells were first cultured in vitro in 1978, facilitating research into the disease.

So what does medical science know about new treatments for this disease now?

Hodgkin's Lymphoma: Aims of Treatment

Hodgkin's lymphoma comprises just 1% of all cancers, but the likelihood of a cure has improved over the past four decades.

Currently, 94% of affected patients are expected to survive. Hodgkin's lymphoma most commonly affects young adults, so oncologists focus on providing a life free of the effects of cancer and free from the effects of the treatments used to cure it. Patients currently experience cancer as a long-term condition – patients may be cured of Hodgkin's lymphoma yet die years later from complications due to chemotherapy.

Diagnosing Hodgkin's Disease

Hodgkin's lymphoma appears as an enlarged lymph node, but can spread to the organs, including the lungs. Symptoms include painless swelling of the lymph nodes in the neck, armpits, or groin, night sweats, weight loss, loss of appetite and general itchiness. Doctors can diagnose Hodgkin's lymphoma with a biopsy of the lymph node – the malignant Reed-Stemberg cells are large, especially in lung cancer, and noncancerous cells usually surround the lymphoma cells.

Hodgkin's lymphoma is curable, but the earlier the disease is diagnosed, the more effective the treatment. When Professor of Oncology Andreas Engert at the University Hospital Cologne, Germany, was asked by Decoded Science about the prospects for those diagnosed quite late with Hodgkin's lymphoma, Professor Engert replied that "*the chances of a full recovery are down to 50%*".

Share

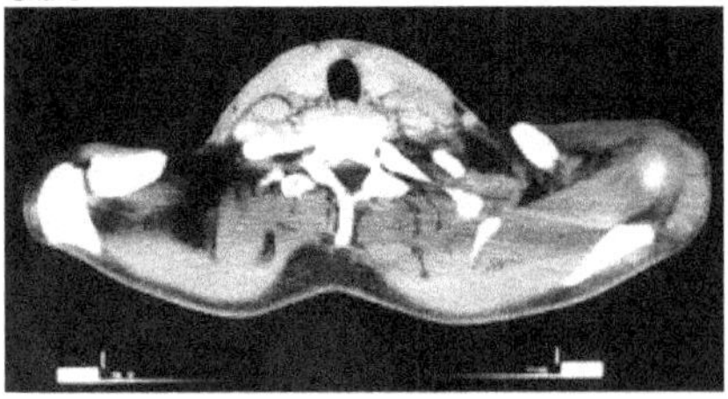

CT scan of Hodgkin's lymphoma in a 46-year-old man with LHS clear lymphoma. Image by J Heuser

Hodgkin's Treatment May Cause Tumours

Standard Hodgkin's treatment includes radiation therapy or chemotherapy such as the combination of drugs known as BEACOPP (Bleomycin, Etoposide, Adriamycin, Cyclophosphamide, Vincristine, Procarbamide and Prednisone) which is 20% better at tumour control and 11% better for overall survival than ABVD (Doxorubicin, Bleomycin, Vinblastine and Dacarbazine). BEACOPP is also far more effective for long-term survival, but according to Simon Crompton at Cancer World, studies have shown that the risk of developing neoplasms, or tumours, after treatment for Hodgkin's lymphoma is 22% at 25 years of age; a far greater risk than experienced by the general population. So the search for alternatives to current chemotherapy and radiotherapy is ongoing.

A New Treatment Regimen for Hodgkin's Disease

During the last three decades monoclonal antibodies have been developed for other cancers, and now there are several new drugs for Hodgkin's lymphoma undergoing trials.

Professor Engert and other researchers are excited about a new targeted drug, brentuximab vedotin SGN-35, an antibody-drug conjugate shown to induce remission in 75% of patients with refractory Hodgkin's lymphoma, a form of the disease that is more severe, with 35% of patients achieving complete remission from the cancer. This particular treatment would avoid the dangers associated with radiotherapy and systemic treatments.

Hodgkin's Lymphoma involves long-term treatment over 20 to 30 years, and the incidences of long-term side effects are not yet known. The best judge of long-term treatments is time, so patients are still monitored long after receiving treatment.

Hodgkin's Lymphoma

Hodgkin's lymphoma is a disease that more commonly affects young people, who then require constant monitoring over the long-term, but have excellent recovery prospects. The treatment of Hodgkin's disease is not straightforward – standard treatments such as BEACOPP are too toxic for normal use, with side effects being a major concern, but there is no room for complacency as secondary cancers can occur within a year due to current treatment regimens. So science continues to pursue new treatments, and medical personnel work to find drugs that can improve long-term recovery rates – such as the targeted drugs sought by Professor Engert.

References

Crompton, S. Andreas Engert: Learning from Hodgkin's. Cancer World. Accessed December 17, 2012.

Immunotherapy: The Exciting Prospect of Harnessing the Body's Immune System to Treat Cancer

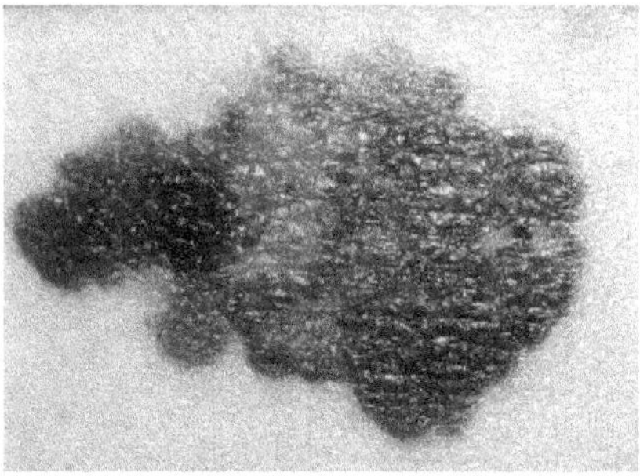

Is immunotherapy the answer to skin cancer? Image by National Cancer Institute

Improved outcomes in cancer patients can be achieved by improving their immune responses to the tumour. People are living longer with cancer, and a diagnosis of cancer in 2013 is not necessarily a death sentence. Here's how immunotherapy is helping patients survive cancer.

What is Immunotherapy?

Immunotherapy, which is also described as biological therapy, harnesses the body's immune system to shrink tumours, control the growth factors that stimulate tumour growth, and aid in the repair of damage caused by other cancer treatments. Early immunotherapy research used T cells to target skin cancer cells, particularly melanoma. The early attempts to improve human immune responses concentrated on the interleukins and interferon. These agents were toxic with poor overall survival rates and this area of research was disappointing.

Immunotherapy has improved as a treatment option due to extensive comparisons to the dynamics of the immune system in chronic infectious diseases, such as tuberculosis (TB) and HIV.

Cancer-Fighting T Cells Can Be Directed Against Tumours

T cells are an additional therapy that can be used with chemotherapy and radiotherapy. Certain T cell populations are specific for tumour cells and used to treat patients with small tumours because of consistently successful outcomes. Fortunately, the immune system has a memory, and an induced response may not only be effective, but can also last for a long period of time.

Trials of immune-stimulatory antibodies targeting cell death proteins have shown promising results in difficult-to-treat cancers, such as melanoma and kidney cancer, as well as certain types of lung cancer. Bristol-Myers Squibb has conducted clinical trials with some of these agents – they also market Yervoy, the brand name for ipilimumab, a treatment for metastatic melanoma.

Immunotherapy Cancer Treatments

There are currently a number of well-established immunotherapy treatments available, including bone marrow and stem cell transplantation for blood disorders. Therapeutic vaccines are more promising, however, because most human cancers have many causal factors. Researchers develop vaccines from a variety of sources, which may reduce their efficacy, and are currently conducting a number of vaccine trials for lung cancer.

Chemotherapy May Induce Immunological Effects

The evident successes of chemotherapy and radiotherapy treatments may be due to immunological factors, not just cell toxicity. Chemotherapy does suppress the immune system, but the immune response to tumour antigens may not decrease, but actually alert the immune system to the cancer.

Immunotherapy and Cancer: Problems There is little correlation between drug exposure and the efficacy and toxicity of immunotherapy. Unexpected side effects, such as kidney failure, may occur and be difficult to treat in a clinical situation, and the high cost of these drug treatments means that patients must be pre-selected and can only be treated if they have pre-existing antibodies. Although immunotherapy shows promise, science is still looking for the ever-elusive cancer-cure.

References

Lesterhuis, W.J., et al. *Cancer Immunotherapy Revisited*. (2011). Nature Reviews Drug Discovery. Accessed February 3, 2013.

MD Becker Partners. *Cancer Immunotherapy Catalysts*. Accessed February 3, 2013.

Skin Cancer Foundation. *New Melanoma Treatment Approved*. (2011). Accessed February 3, 2013.

Beishon, M. *We told you so: How the persistence of immunotherapy researchers is finally paying off.* (2012). Cancerworld. Accessed February 3, 2013.

Stop Cancer Now! World Oncology Forum Effort to Reduce Cancer by 25% by 2025.

Delegates attending the World Oncology Forum in Lugano, Switzerland, October 2012. Image by 2012 HarrisDPI

The World Health Organization's prediction that by 2030, 22 million men, women and children will be diagnosed with cancer every year, and 13 million will die from the disease, prompted cancer experts on February 4, 2013, World Cancer Day, to call on governments to halt the dramatic worldwide increase in deaths from cancer using a widely published appeal called Stop Cancer Now!

Governments must deliver on commitments made at the World Health Assembly in May 2012 to cut deaths from cancer by 25% by 2025. The deaths due to cancer are estimated to cost $900 billion annually.

Stop Cancer Now!

The World Oncology Forum (WOF), organized by the European School of Oncology in partnership with *The Lancet*, is determined to monitor the fight against cancer. WOF members met in October 2012 and called on world leaders to introduce strategies similar to the international action used to counter AIDS in 1993. Cancer is a major cause of death globally, with the rate of new cases expected to double over 25 years.

Meeting the commitment for reducing cancer deaths could save 1.5 million lives worldwide each year. The STOP CANCER NOW! appeal in February 2013 warned that current strategies for controlling cancer are not working. This appeal appeared in the *International Herald Tribune*, *Le Monde*, *El País*, *La Repubblica* and *Neue Zürcher Zeitung,* along with Articles in *The Lancet* and Cancerworld magazine.

Cancer Community Alarmed at Escalating Crisis

Cancer specialists question the lack of urgency over cancer deaths, particularly when the AIDS epidemic was forced onto the agenda of the G8 summits. When will tackling cancer receive the same attention? International cancer experts from all over the world met at the WOF in Lugano, Switzerland, in October 2012, which included health journalists and, in particular, Rifat Atun, who directed the Global Fund to Fight AIDS, Tuberculosis and Malaria. Atun led the discussion on how the international effort against cancer could learn from that experience. Richard Horton, editor of *The Lancet*, was there to ensure that a working strategy with clear objectives was set up to turn the tide on cancer.

Cancer is More Common Worldwide

In the developed world, people are exercising less, eating less healthily, and drinking more alcohol. This results in yet more cancers, yet the death rates from cancer are falling. Anna Wagstaff at CancerWorld pointed out the work of Dr Richard Peto from the University of Oxford, who described all the reductions in death from cancer as being due to decreases in smoking-related cancers from 1965-2010. The death risk from all other cancers remained the same in 2010. One delegate at the WOF stated that concerted action is needed to deal with tobacco, particularly in India and China.

Cancer: Investing in Prevention

Wagstaff also pointed out the work of Paolo Vineis, Chair of Environmental Epidemiology at Imperial College in London, who examined the evidence for the causes of cancer and how it can be prevented. Most cancers are due to environmental and lifestyle factors interacting with genetic susceptibility. Professor Vineis stated that cancers caused by infectious diseases, such as liver and cervical cancer, offer tremendous opportunities for prevention – and vaccination is intended to control these two cancers. This a key strategy on which the WOF participants agree. The group's 10-point strategy to counter cancer was published as the Stop Cancer Now! appeal in February 2013. Richard Horton echoed one advocate's statement that "*we need to mobilise the cancer community and engage people into forcing governments to act.*"

Cancer Treatments: The Future

The goal of ensuring every cancer patient has access to diagnostics and curative and palliative care is attainable. This must be central to the international strategy for dealing with cancer.

References

Wagstaff, A. *Stop Cancer Now!* Cancer World. Accessed February 15, 2013

European School of Oncology. *World Oncology Forum*. Accessed February 15, 2013

World Health Organization. *Cancer Statistics*. (2013). Accessed February 15, 2013

European School of Oncology. *Stop Cancer Now! Appeal*. (2013). Accessed February 15, 2013

European School of Oncology. *Stop Cancer Now!* (2013). Accessed February 15, 2013

Researchers Discover how Staphylococcus Aureus, Colonizing the Human Nose, Spreads MRSA

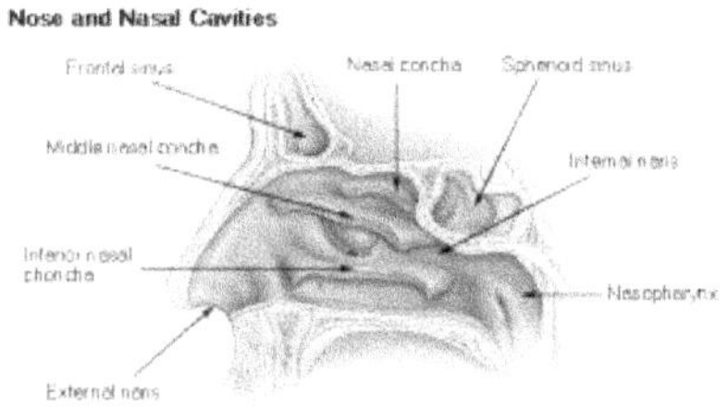

S. aureus colonises the nose and nasal cavities. Image by National Cancer Institute

Could new research could reduce MRSA infections in hospital patients?

Researchers from the Department of Biochemistry and Immunology and Department of Molecular Microbiology at Trinity College Dublin in Ireland have identified how the bacterium *Staphylococcus aureus* colonizes human nasal passages. *S. aureus* is the precursor form to the antibiotic resistant bacterium MRSA, and the presence of the bacteria in the nose makes it a risk factor for infection.

How *S. aureus* Colonises the Human Nose

S. aureus has the potential to cause severe invasive disease and is a major concern in hospitals and healthcare facilities, where infections are sometimes caused by antibiotic resistant strains such as MRSA (methicillin-resistant *S. aureus*). *S. aureus* colonizes at least 20% of the human population by binding to cells within the nasal cavity, which may cause an infection. A recent study, published in the Journal *PLoS Pathogens*, shows that a protein located on the bacterial surface, called clumping factor B (ClfB), recognizes a protein called loricrin – a major component of the cells inside the nose.

Bacterial Clumping Factor Binding to Loricrin Facilitates *S. aureus* Colonisation

Previous studies have shown that ClfB encourages *S. aureus* colonisation of the nasal cavities. This most recent study, however, actually identifies the mechanism by which ClfB encourages *S. aureus* nasal coloniation.

The study found that ClfB binding to a protein called loricrin was crucial for successful colonization of the nose in mice, and that fewer bacterial cells colonized the nasal passages of a mouse lacking loricrin, compared to a normal mouse. When *S. aureus* strains lacking ClfB were used, nasal colonization was dramatically reduced, and nasal administration of loricrin reduced S. aureus colonization of the mice.

The results suggest that there is a way to reduce colonisation, and disease risk in the process, but only if these results can be translated to humans.

Dock, Lock and Latch, and Mupirocin

The binding of ClfB to loricrin was also found to be crucial for *S. aureus* colonization in the human nose; soluble loricrin reduced the binding of *S. aureus* to human nasal skin cells. The researchers also found that the dock, lock and latch mechanism, a specific interaction between the proteins caused by a change in their structures as they interact, results in firm binding of *S. aureus* to the cells of the human nose.

We currently use the antibiotic mupirocin to reduce bacterial colonization in the nasal cavities, but need to review its use due to problems with antibiotic resistance. Understanding the molecular interactions and mechanisms of nasal colonization, such as dock, lock and latch, can help find new treatments.

Finding New Antibiotic Treatments

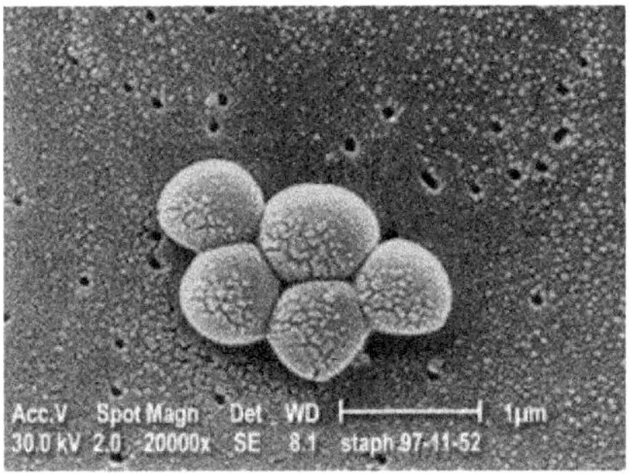

MRSA bacteria can cause fatal infections. Image courtesy of Keith Gregg – Curtin University

Assistant Professor Rachel McLoughlin and Professor Tim Foster, the study's corresponding authors, concluded: "*Loricrin is a major determinant of S. aureus nasal colonisation. This discovery opens new avenues for developing therapeutic strategies to reduce the burden of nasal carriage and consequently infections with this bacterium. This is particularly important given the difficulties associated with treating MRSA infections.* "Dr. McLoughlin told Decoded Science how this research could reduce the incidence of MRSA in hospitals, where postoperative patients are particularly vulnerable: "*Nasal colonisation with S. aureus has been identified as a clear risk factor for subsequent invasive infection. If someone goes into hospital for surgery and they are already carrying S. aureus in their noses, then, they are at an increased risk of getting an infection, e.g in their surgical wound, with the same strain they harbor in their noses. S. aureus is transferred through skin-to-skin contact. By identifying a mechanism to irradiate S.aureus from the noses this could be used to decolonize patients before they go into hospital.*"

Dr. McLoughlin reiterated that MRSA is a specific antibiotic resistant strain of *S. aureus*, and only "*a very small % of the general population are actually colonised with MRSA in their noses.*" But she cautioned, "*any infection with S. aureus had the potential to mutate into an MRSA strain particularly in the hospital setting where there is selective pressure from high use of antibiotics.*"

References

- Mulcahy, Michelle E., et al. *Nasal Colonisation by Staphylococcus aureus Depends upon Clumping Factor B Binding to the Squamous Epithelial Cell Envelope Protein Loricrin*. (2012). PLoS Pathogens. doi:10.1371/journal.ppat.1003092. Accessed February 22,

An Infection with Hepatitis C May Eventually Lead to Liver Cancer

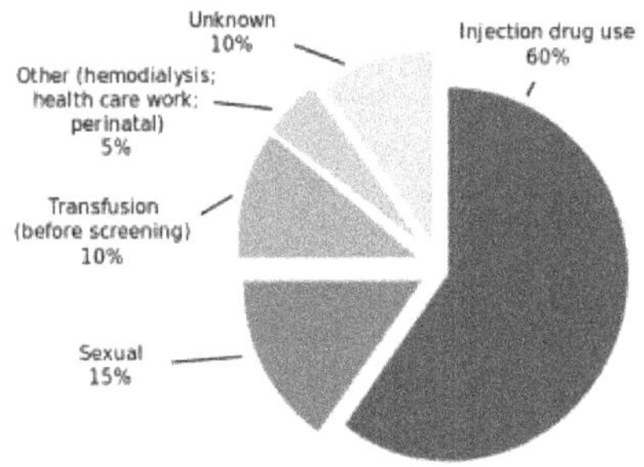

Injection drug use is the single most common source of hepatitis C infection. Image by Optigan13

The hepatitis C virus causes a contagious liver disease by the same name, resulting in lifelong illness. Hepatitis C virus infection occurs when the blood from an infected person transfers to another person. 3–4 million hepatitis C virus infections occur annually; 150 million people worldwide have chronic hepatitis C infections and are at risk of developing liver cancer. At least 350,000 people die from hepatitis C-related liver disease each year.

Hepatitis C Disease

Hepatitis C incubates for up to 6 months, and at least 80% of infected people lack any symptoms. The few people who are acutely symptomatic present with fatigue, nausea, joint pain and jaundice, which is a yellowing of the skin and eyes. A total of 85% of newly-infected persons develop chronic hepatitis, and 70% of chronically infected people develop chronic liver disease, including cirrhosis (5-20%). A total of 1–5% of people with chronic hepatitis C die from cirrhosis or liver cancer. In fact, the underlying cause of cancer in 25% of liver cancer patients is hepatitis C.

Hepatitis C Infection: Diagnosis

Doctors sometimes miss hepatitis C diagnoses because most people infected with the virus lack symptoms. The presence of antibodies against the hepatitis C virus indicates that a person is or was infected by the virus. The presence of antibodies to the hepatitis C virus in the blood for longer than 6 months is confirmation of the diagnosis. Specialized tests evaluate patients for liver disease, including cirrhosis and liver cancer.

One Patient's Experience of Long-term Hepatitis C

An 88-year-old, otherwise healthy man, received a diagnosis of liver cancer in October 2012, and with hepatitis C in February 2013. As of March 2013, he has slow-growing liver cancer in one lobe. He never received a blood transfusion, has no pain, and exhibits no signs or symptoms of liver cancer. He represents the 10% of liver cancer cases with unknown aetiology.

Janssen Pharmaceutica Outlines New Anti-Viral Treatments for Hepatitis C

Decoded Science asked Janssen Pharmaceutica if their new anti-viral treatments are effective for long-term hepatitis C sufferers: *"As the number of available treatments for hepatitis C has increased, so has the chance of being cured. Changes in treatment over time have led to an increased chance of SVR."* A sustained virologic response (SRV) is a lack of virus in the blood 24 hours, called aviremia, after treatment with new antiviral agents.

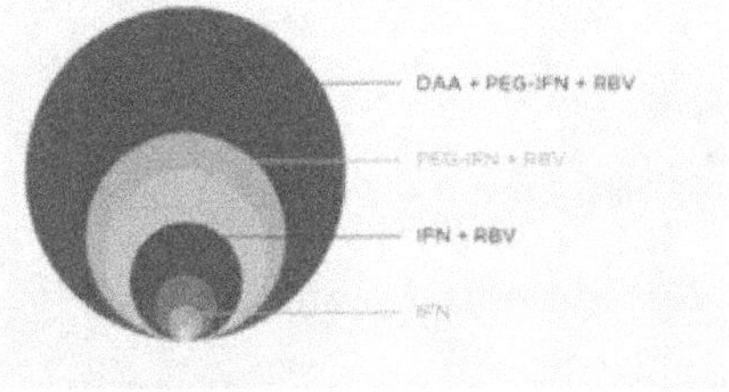

Treatment regimens for hepatitis C. Image by Janssen Pharmaceutica

"*Protease inhibitors, also known as Direct Acting Anti-Virals* [DAA], *help to prevent the hepatitis C virus from multiplying. They have been shown to increase the chance of a cure for all patients when taken in combination with pegylated interferon alfa* [PEG-IFN] *and ribavirin* [RBV] *versus pegylated interferon alfa and ribavirin alone. Currently there are two available protease inhibitors: Boceprevir and Telaprevir.*"

Blood Transfusion Screening

Hospitals did not screen blood transfusions for hepatitis C prior to the 1970s, so blood transfusions prior to that era are a risk for long-term hepatitis C infection. The consequences of long-term infection with hepatitis C are now apparent, with elderly patients requiring nursing home care. Some countries make provisions for the lack of screening by paying for the nursing home care for those who are infected, but we do not know the true number of people infected with hepatitis C worldwide.

References

World Health Organization. *Hepatitis C*. (2012). Accessed March 15, 2013.

Stop Hep C. *Case Studies*. Janssen Pharmaceutica. Accessed March 15, 2013.

Hansen, P and Flora, K. *FAQ: The link between hepatitis C and liver cancer*. Providence Health & Services. Accessed March 15, 2013.

Antibiotic Resistant Bacteria may Lead to Intractable Gonorrhoea Infections and Restrict Operations

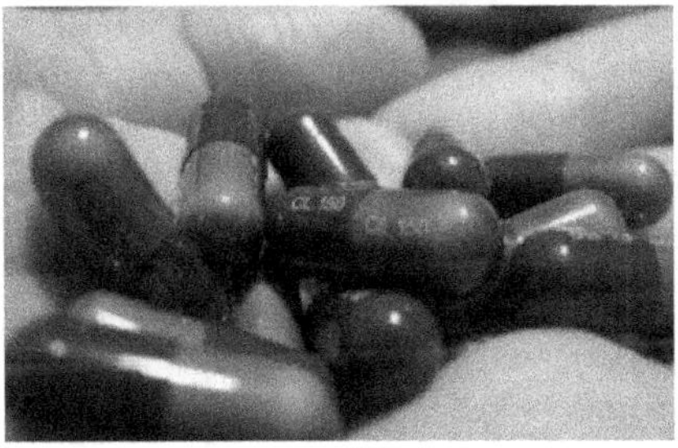

Misuse of antibiotics is leading to resistance. Photo by lamentables

Antibiotics have been available to doctors for at least 70 years, allowing for sterile surgical procedures. But the misuse of antibiotics by patients and farmers is leading to the problem of antimicrobial resistance (AMR). Some infections that were thought to be under control, such as gonorrhoea, can now present in patients as serious infections resistant to treatment with antibiotics.

Antimicrobial Resistance

Microbial resistance to the beta-lactamase ring found in many common antibiotics have emerged among some gram-negative bacilli, rendering powerful antibiotics ineffective at treating infections. Antimicrobial Resistance, or AMR, is a serious problem with *Neisseria gonorrhoeae*, the infectious agent that causes gonorrhoea, reversing gains made by doctors in the control of this common sexually transmitted infection. Patients must now take oral cephalosporins to control gonococcal infections, and as a consequence, the prevalence of the infection is increasing worldwide; untreatable gonococcal infections could result in more deaths.

What Encourages Antimicrobial Resistance?

The emergence of AMR is a complex problem driven by many interconnected factors. Inappropriate and irrational use of antimicrobial drugs, such as antibiotics, provide favourable conditions for the emergence and spread of resistant microorganisms. An example is when patients do not complete the full course of a prescribed antibiotic or when they use poor quality antimicrobials.

Another inappropriate use of antimicrobials is the widespread unrestricted use of antibiotics in farm animals, resulting in antimicrobials entering the normal food supply. Antibiotics that farmers previously used to promote growth in beef cattle have found their way into wildlife, with methicillin-resistant *Staphylococcus aureus* (MRSA) becoming more common. Antibiotic residues are also present in the water supply, often in the presence of the common contaminant *Escherichia coli*.

An inadequate response of officials to the uncontrolled spread of disease and insufficient engagement of community resources certainly aids the spread of infection. Poor quality medicines and an inadequate supply of medicines are available to counter diseases in Africa, where patients purchase antibiotics wholesale. Poor infection prevention and control practices, and insufficient research and development of new products in modern medical microbiology also encourages AMR. Single, isolated interventions are of little use to patients.

Pharmaceutical Companies Withdraw from Antibiotic Development

Pfizer, Roche, Bristol-Myers Squibb and Eli Lilly have reduced their antibiotic research efforts, although Merck & Co. and Glaxo-Smith-Kline are still actively pursuing antibiotic research. Due to the reduction in research efforts, few classes of antibiotics have been developed since 1987.

Dr. Gary Roselle, National Director, Infectious Diseases Service for the Department of Veterans Affairs health care system, told The Washington Times, "*None of them really is going to be active against these bacteria in the near term.*"

Combating the Threat

A global and national multisectoral response is urgently needed to combat the growing threat of AMR. In the absence of a global response, mankind may lose the battle in the human body, whose cells are outnumbered by beneficial commensal bacteria.

References

Wetzstein, C. *Antibiotic-resistant 'superbugs' alarm health care industry*. (2013). The Washington Times. Accessed March 27, 2013.

Collignon, P. *We can beat superbugs – here's how*. (2012). The Conversation. Accessed March 15 2013.

New Anthrax-Killing Antibiotic Also Eliminates MRSA

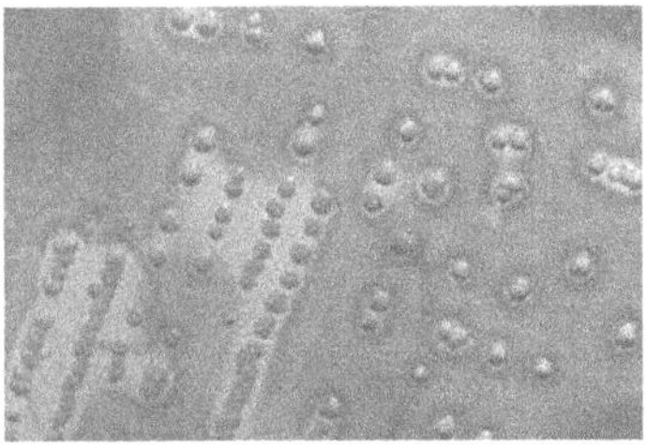

Staphylococcus aureus on blood agar plate showing haemolysis. Image by Microrao

A new broad-range antibiotic kills a wide range of bacteria, including intractable Methicillin-Resistant *Staphylococcus aureus* (MRSA) infections which fail to respond to traditional antibiotics, and the Anthrax bacteria.

Scientists at Rockefeller University and Astex Pharmaceuticals jointly developed the new antibiotic, Epimerox, which uses weak spots in the bacterial cell walls as a point of attack. These weak spots are already a target for bacteriophage viruses, AKA 'phage' viruses, which infect and replicate inside bacteria.

Viral Tactics for Infecting Cells

Bacteriophage infect bacteria by targeting weaknesses in their cell walls, and this new antibiotic uses the same successful tactic. *"We're taking advantage of what phage have 'learned' during this period for us to identify new antibiotic targets we believe will escape the problem of resistance found for other antibiotics,"* says senior author Vincent Fischetti, head of the Laboratory of Bacterial Pathogenesis and Immunology, Rockefeller University, New York.

Fischetti and his colleagues used a molecule encoded with phage genes to identify a bacterial target enzyme called 2-epimerase, which is not native to the human body – an amino acid in the cell wall that dissolves in water.

In 2008, Fischetti's lab, along with Rockefeller University's Erec Stebbins and his colleagues in the Laboratory of Structural Microbiology, mapped the internal crystal structure of the 2-epimerase enzyme. The researchers identified a previously-unknown regulatory mechanism in 2-epimerase, and targeted it when developing a new multi-use antibiotic.

Raymond Schuch, a former postdoctoral researcher in Fischetti's lab, tested the new antibiotic in mice infected with *Bacillus anthracis;* Epimerox fully protected the mice from anthrax bacteria. Further, the bacteria did not develop resistance to the inhibitor over the course of the research. The researchers also found that Epimerox was able to kill Methicillin Resistant *Staphylococcus aureus* (or MRSA) with no evidence of resistance developing, even after extensive testing.

"Since nearly all Gram-positive bacteria contain 2-epimerase, we believe that Epimerox should be an effective broad-range antibiotic agent," says Fischetti. *"The long-term evolutionary interaction between phage and bacteria has allowed us to identify targets that bacteria cannot easily change or circumvent. That finding gives us confidence that the probability for developing resistance to Epimerox is rather low, thereby, enabling treatment of infections caused by multi-drug-resistant bacteria such as MRSA.* A Universal Antibiotic?

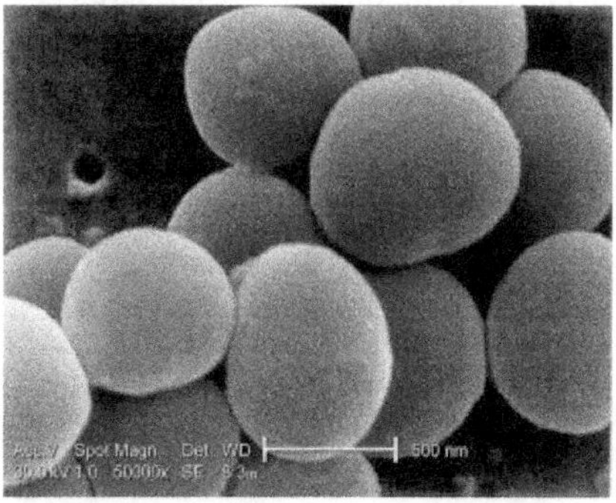

Could a newly-developed antibiotic fight deadly hospital-acquired MRSA infections? Image courtesy of the US CDC.

Epimerox was specifically designed to fight against *Bacillus anthracis* 2-epimerase, however, it also inhibited the growth of many Gram-positive (but not Gram-negative) organisms that encode 2-epimerases, including MRSA. (A Gram-positive bacteria contains peptidoglycan and turns dark blue or violet during Gram Staining, while Gram-negative bacteria turn pink or red.)

For example, the presence of this new antibiotic was enough to completely stop the growth of *Staphylococcus aureus* strain RN4220 bacteria over a 12 hour period.

MRSA a Major Health Risk

MRSA infections have even been detected in wildlife, which fully confirms its prevalence in multiple populations. Currently, the antibiotic vancomycin is used as a last resort in MRSA infections.

When asked by Decoded Science if Epimerox could be the definitive antibiotic for MRSA, Dr. Fischetti replied: '*Right now VancomycinRSA organisms are being isolated, so I believe that by the time Epimerox hits the market (in 5-6 yrs), these organisms will be prevalent and Epimerox will be able to control them and other resistant staphylococci.*'

In other words, Epimerox could very well prevent and successfully treat intractable and life-threatening MRSA infections in hospitals.

References

Raymond Schuch, Adam J. Pelzek, Assaf Raz, Chad W. Euler, Patricia A. Ryan, Benjamin Y. Winer, Andrew Farnsworth, Shyam S. Bhaskaran, C. Erec Stebbins, Yong Xu, Adrienne Clifford, David J. Bearss, Hariprasad Vankayalapati, Allan R. Goldberg, Vincent A. Fischetti. *Use of a Bacteriophage Lysin to Identify a Novel Target for Antimicrobial Development*. (2013). PLoS ONE. Accessed April 14, 2013.Shylo E. Wardyn1,2, Lin K. Kauffman3 and Tara C. Smith. *Methicillin-resistant Staphylococcus aureus in Central Iowa Wildlife*. (2012). Journal of Wildlife Diseases. Accessed April 14, 2013.

Measles Outbreak in U.K. Linked to Poor MMR Vaccine Uptake After Autism Scare

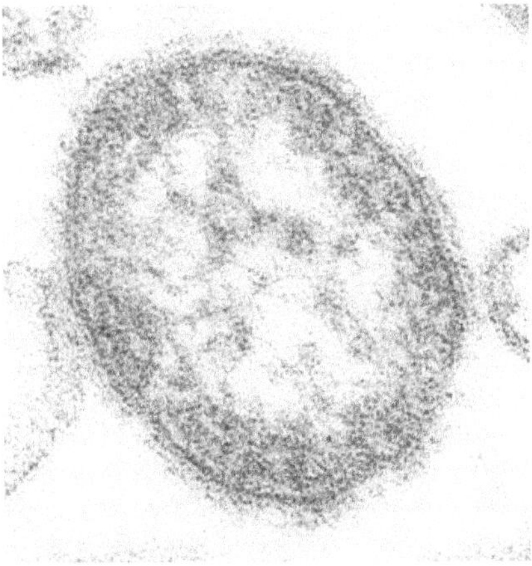

The measles virus transmits rapidly through high-population areas. Image courtesy of the U.S. CDC

As of April 5, 2013, there were 942 cases of measles in South Wales, UK, and Health Officials say there's no way of knowing if they have reached a peak in the outbreak. Where did all these cases of measles come from?

In 2003, Dr. Andrew Wakefield's research paper revealed a link between the MMR (Measles Mumps Rubella) and autism which resulted in a 20% reduction in vaccination in England and Wales. Dr. Wakefield's research has since been discredited, but the reduction in vaccination rates continues, with dangerous results.

In England in 2012, there were 202 cases of measles – and there have been 887 cases of measles so far in 2013. The British government is due to spend £20 million vaccinating 1 million 12-16 year old schoolchildren within the next month, in hopes of controlling the measles epidemic.

Measles in Isolated and Urban Populations in the UK

Measles is occurring in isolated populations with traditionally low vaccination rates, along with populations of kids whose parents decided against vaccination based on the autism study. According to The Guardian, there have been 81 measles cases among itinerants and 41 measles cases among the Orthodox Jewish community in the UK. School outbreaks affect less than 1% of children, however they are of concern due to the potential spread. London itself has had 67 confirmed measles cases in 2013. Since the MMR autism scare arose in 1998, few parents in London boroughs took their children for vaccination; parents were suspicious of government reassurances. *'I worry about London,'* said Professor David Salisbury, director of immunization at the UK Department of Health, according to The Guardian. *'It is a fast-moving group of people, with families coming in and moving out. Historically there is a legacy of poorer immunisation. London to its credit has done a great deal of good work to pull up immunisation coverage, but that is for the young children. People are densely packed together in London and that's just what measles like for higher levels of transmission.'*

Autism/Vaccination Claim

According to The Guardian, Public Health Wales, UK, the body coordinating the response to the measles epidemic, has demanded a private health centre remove claims on its website that their single shot is "*more effective than the MMR*" - and has asked the site to remove links to Dr. Wakefield's claim of a link between the MMR vaccine and autism.Measles is a serious life-threatening disease. The 2013 UK epidemic reinforces the fact that complacency in vaccination uptake from a decade before could have serious repercussions for children living in a high density population area like London or South Wales. There is no equivalent measles outbreak in the United States as yet, although 2012 saw unusually high rates of chicken pox as a result of anti-vaccine controversies and concerns that mercury-containing vaccines cause autism.

References

Centre for Disease Control facts. *CDC Vaccines MMR*. (2013). Accessed April 28, 2013

The Guardian Measles Outbreak. *Guardian MMR Vaccinations*. (2013). Accessed April 28, 2013

McWatt, J. *Measles Latest: Epidemic May Not Peak for Weeks*. (2013). Wales Online. Accessed April 28, 2013.

Williams, A. *Measles threat 'could spread across the UK': Top doctor fears Welsh outbreak may grow*. (2013). UK Daily Mail. Accessed April 28, 2013.

Boseley, S. *Lancet Retracts "Utterly False" MMR Paper*. (2010). Accessed April 26, 2013.

Lyme Disease: Tick-Borne Encephalitides and Complications

Adult Deer Tick, Ixodes scapularis, Photo Scott Bauer, US Dept of Agriculture

Lyme disease is the world's fastest-growing tick-borne infection. Victims can have either mild or severe symptoms, depending on the state of their immune system and the extent of co-infections in the body such as bartonella, anaplasmosis and babesia or other health issues. When do Lyme Disease symptoms show up, and what makes them worse?

Lyme disease (Borreliosis) Symptom Onset

Some who contract Lyme Disease may have no symptoms at first; they can appear months after the initial infection. You can catch European Lyme disease in forested regions throughout Europe and northern Asia; it's more common in eastern and central Europe. In North America, several species of *Borrelia burgdorferi* cause Lyme disease, all over the continent. Infection by American ticks results in similar symptoms, although U.S. tests cannot detect European Lyme disease.

Lyme Disease and Co-Infections

The bacterium *Borrelia burgdorferi* causes Lyme Disease, which we get when we're bitten by infected ticks. If left untreated, infection can spread to joints, the heart, and the nervous system. Doctors diagnose Lyme disease based on symptoms and exposure to infected ticks, and lab testing. Once you've received a diagnosis, your doctor will treat you with antibiotics.

There are a variety of co-infections that can make Lyme Disease even worse. If you've got any of these infections, and add an infected tick to the mix, you could become very ill.

- *Bartonella* bacteria cause cat scratch disease, trench fever and Carrión's disease.
- The bacterium *Anaplasma phagocytophilium,* carried by the black-legged tick (*Ixodes scapularis*) causes Anaplasmosis.
- Microscopic *Babesia* parasites cause Babesiosis, which infects red blood cells and is also spread by ticks.

Potentially Fatal Tick-Borne Encephalitides

Flaviviruses transmitted by ticks cause encephalitis. The Central European form of encephalitis has two phases, the first phase beginning with a fever which lasts for 7 days, followed by an asymptomatic period of 8 to 15 days. The second phase manifests with the signs and symptoms of meningo-encephalitis, including fever, headache, nausea and vomiting; body temperature may rise rapidly. Most patients recover completely, but a lack of treatment can be very dangerous.

A small, almost invisible tick bite could prove fatal to you, for example, if you've had your spleen removed. A 2008 case in the Journal of American Board of Family Medicine highlights an incident in which a traveller did not reveal a visit to a sector of the United States where tick borne encephalitis is endemic – his trip could have proved fatal.

Lyme disease ticks may also transmit other dangerous tick-borne diseases:

- **Imported tick-borne spotted fevers** (rickettsial infections) cause infection in returning travellers. In the U.S., *Rickettsia africae* (the agent of African spotted fever) causes the most frequently-diagnosed infection due international travel.
- **Tick-borne encephalitis** is present in many parts of Europe, the former Soviet Union, and Asia. Three virus sub-types include: European or Western tick-borne encephalitis virus, Siberian tick-

borne encephalitis virus, and Russian Far Eastern tick-borne encephalitis virus.

Tick-Bite Season is Here: Lyme Disease

Another traveller, an orienteer who competed in Central Europe, returned home with Central European meningo-encephalitis which was diagnosed and treated- fortunately without complications. He told me, '*A bull's eye, spreading rash with the tick bite clearly visible coupled with a rise in body temperature ensured a visit to my GP with a seven day course of antibiotics clearing the rash and infection.*'

Red Deer, coming from the Czech Republic, have helped spread tick-borne encephalitis throughout Western Europe. Alarmingly, tick-borne encephalitis is on the increase, so this traveller may not be the last to experience the bull's-eye rash of Lyme Disease.

References

US CDC. *Lyme Disease Centre*. (2013). Accessed June 2, 2013.

Abrams, Y. *Complications of Babesia and Lyme Disease*. (2008). Journal of American Board of Family Medicine. Accessed June 2, 2013.

National Science Foundation. *Lyme Disease on the Rise*. Accessed June 2, 2013.

Russian Prisoners with Multi-Drug-Resistant Tuberculosis are a Threat to AIDS/HIV Patients Worldwide

Can prisoners become sick from staying in unsanitary prison conditions? Image by kconnors

Russian officials arrested and imprisoned 30 Greenpeace activists who climbed on a GazProm oil rig in the Barents Sea, August 2013.

The activists, after being released, February 2014, as part of President Putin's attempt to showcase Russia at the Sochi Winter Olympics, immediately reported on the horrific conditions of their detention facilities to world media.

Russian Prisons: a Reservoir of Infection

Russia has a reservoir of extensive, long-term tuberculosis infections. Tuberculosis spreads in the community as prisoners leave, as well as through those who work at the prison, including medical personnel.

Unfortunately, due to globalization, tuberculosis is not just a significant public health concern for Russia – these infections could spread anywhere, thanks to our interconnected intercontinental air travel system.

Tuberculosis takes advantage of compromised immune systems. Large populations, poor nutrition, high stress, and poor living conditions are all contributing factors, and all of these conditions are present in the Russian prison system. In addition, prisoner quarantine keeps the infected people together, which results in re-infection.

In the non-immune (susceptible) host, the bacilli initially multiply unopposed by normal host defence mechanisms, including the macrophage during phagocytosis (when the immune system cell engulfs another cell) and may remain viable for extended periods of time.

The Spread of Tuberculosis

Tuberculosis spreads when an infected person sneezes or coughs. *Mycobacterium tuberculosis* remains in the air for days and may travel great distances – so once an outbreak starts, it tends to last a long time. Very few tuberculosis cells are necessary for the infection to reach the lungs – which provide a high concentration of oxygen and helps tuberculosis survive.

One person infected with tuberculosis may infect up to 15 people. Family household contacts, especially children in schools who are clearly not immuno-compromised, along with individuals working or living in enclosed environments (hospitals, nursing homes and prisons) with an infected person are at major risk of TB infection.

Now, however, some good may come out of new drug treatments for multi-drug-resistant TB.

Médecins Sans Frontières

According to a 2004 study, published in the Lancet, by Dominique Lafontaine, et al, Médecins Sans Frontières (MSF) started working in Siberian prisons in 1996, where they treated at least 10,000 patients for TB, with the approval of the penal authorities, using the WHO-led DOTS TB treatment strategy. Unfortunately, MSF withdrew from the region in September 2002.

During that six year period, doctors treated many patients with multi-drug-resistant tuberculosis. In Siberian prisons, at least 22% of new cases, and at most 40% of re-treatment cases, were multi-drug-resistant – these are some of the highest multi-drug-resistant TB rates recorded around the world.

MSF applied to treat multiple-drug-resistant TB patients to the Green Light Committee International Authority, using caproemycin, cycloserine and flouoro-quinolones. These antibiotic treatments are reserved for cases involving drug-resistant organisms, treatment failures, extra-pulmonary TB, drug toxicity or patient intolerance to all other agents.

Although the Green Light Committee International Authority approved their application to start using second-line drugs to treat patients in Siberian prisons, the Russian Ministry of Health rejected this application to treat the patients. According to the MSF study, their application was rejected, *"on the grounds that the treatment schemes proposed contradicted the regulations of the Russian Pharmaceutical Committee. It therefore classified the DOTS-Plus pilot project as "experimental", which is forbidden within the penal system under national law."*

Existing Russian drug legislation, imposed by ultra Russian nationalist, President Vladimir Putin, ensured MSF were required to use a Russia-dictated treatment strategy for multiple-drug-resistant tuberculosis. The Russia-approved treatment contradicted the basic treatment principles that the World Health Organization outlines for multi-drug resistant TB. These treatments were crucial for the program to cure recalcitrant TB patients. Regrettably, in September, 2003, MSF pulled their teams out of Siberia and closed down the TB treatment program.

Tuberculosis, Russian Prisons, and Immune-compromised Patients Worldwide

MSF efforts in Russian prisons for six years led to a crisis whereby now, Russian Ministry of Health officials are making concerted efforts to treat multiple-drug-resistant tuberculosis in prisons. Released prisoners harbouring latent tuberculosis could threaten AIDS/HIV patients worldwide. Russian prison staff including prison officers, nurses, doctors and inmates regularly contract TB in the course of their working lives.

What's the answer to this ongoing problem? Application of WHO standards of care in Russian prisons could make all the difference.

Resources for this article

Lafontaine, Dominique. *Treatment of multidrug-resistant tuberculosis in Russian prisons*. (2004). Lancet. Accessed on May 06, 2014

Naidoo, K. *Gazprom's over-reaction to Arctic oil protest is a sign their fortune is at stake*. (2013). The Guardian. Accessed on May 06, 2014

Ahlberg, L. *Multitarget TB drug could treat other diseases, evade resistance*. (2014). University of Illinois. Accessed on May 06, 2014

World Health Organization. *The five elements of DOTS*. Accessed on May 06, 2014

World Health Organization. *The Green Light Committee (GLC) Initiative*. Accessed on May 06, 2014

Scientific American. *Multi-drug Resistant Tuberculosis in Russia*. (2009). Accessed on May 06, 2014

Coker, R. *Risk factors for pulmonary tuberculosis in Russia: case-control study*. (2006). BMJ. Accessed on May 06, 2014

A New Bubonic Plague Outbreak is Forecast

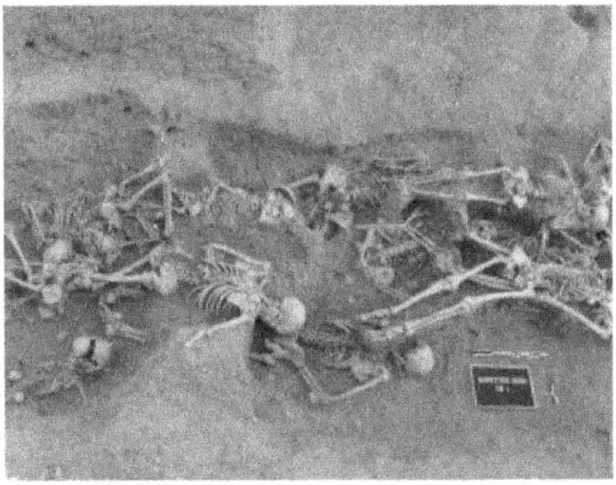

Thousands of individuals died collectively from powerful Black Death epidemics, resulting in multiple mass graves and landfill-type burials. Photo by S. Tzortzis, image by 7mike5000.

Could the world suffer from a new bubonic plague epidemic? If the plague mutated so that current medicines couldn't combat it, we might see an outbreak to rival the Black Death in the 1300s.

Bubonic Plague History

The European rat-borne bubonic plague whose fleas harboured *Yersinia pestis* killed 350 million people in Europe between September 1348 and June 1349, shortly after the European-wide 100 Years War. These deaths represented 50% of the human population in Europe.

The Black Death received its name after the colour of the buboes, or swellings, which disfigured the sick (see image below). The survivors of the European outbreak, who were immune to the plague, re-generated the current highly-successful human population in the region.

Researchers Stephanie Hansch and Barbara Bramanti discovered the route of bubonic plague infection led from Asia to Marseille by November 1347, through western France to northern France and over to England.

However, a different type of *Yersinia pestis* was found in the Netherlands. The researchers believe the infection in the Netherlands came from the North, indicating another infection route, from Norway to the Netherlands. '*The history of this pandemic, is more complicated than previously thought,"* Haensch stated.

Yersinia Pestis Bacteria

Yersinia bacteria have many sub-species, some of which are harmful. *Yersinia pestis*, the bacterium responsible for the bubonic plague or the Black Death, and *Yersinia enterocolitica*, a major cause of gastroenteritis, are the most lethal sub species. The evolution of these *Yersinia* species is compelling.

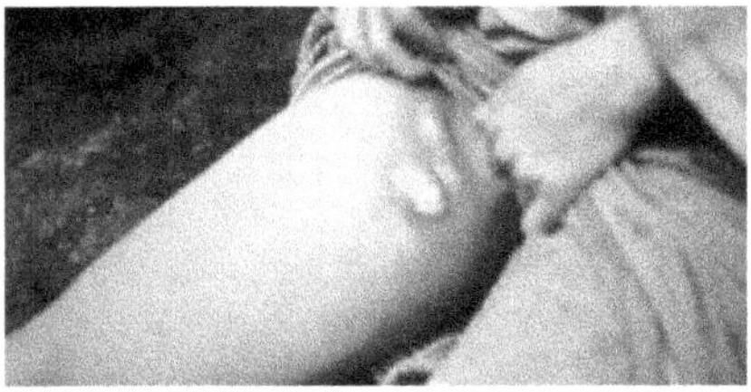

Bubos, on a child's leg in Madagascar, 2014, are the first sign of a Bubonic Plague Infection. Photo courtesy of the CDC, image by Optigan13.

New Bubonic Plague Forecast

Medical researchers extracted DNA from the teeth of victims of a bubonic plague pandemic that swept through the Byzantine Empire in AD 541-542; exhuming the victims from an early medieval cemetery in Bavaria, Germany. The AD 541-542 plague wiped out half the world's population at that time.

Research links the AD 541-542 plague with the Black Death bubonic plague, which rats spread during the 14th to 17th centuries, and in the 19th and 20th centuries.

Yersina pestis bacterium caused all three pandemics, but they weren't quite the same.

Significantly, the bacterial strains of the first pandemic are very different from those of the later pandemics so that scientists warn that the same bacteria with different DNA lineages is a cause for concern.

"These results show that rodent species worldwide represent important reservoirs for the repeated emergence of diverse lineages of Y pestis into human populations," the researchers conclude.

The Future of the Bubonic Plague

Modern-day antibiotics halt currently known strains of plague, yet, researchers warn about potentially dangerous mutations. Should an airborne mutant version emerge, the plague could kill people within 24 hours of infection, cautioned Hendrik Poinar, director of the Ancient DNA Centre at McMaster University in Canada.

With the widespread presence of rodents and the continued existence of evolving strains of plague, experts believe that another outbreak of bubonic plague is possible.

Resources for this article

Hansch, Stephanie, et al. *Yersinia pestis bacteria clearly identified as the cause of the big plague epidemic of the Middle Ages*. (2010). Johannes Gutenberg-Universität Mainz. Accessed on June 22, 2014

Dugdale, David. *Plague: a severe and potentially deadly bacterial infection*. (2011). Medline Plus. Accessed on June 22, 2014

The Lancet. *Yersinia pestis and the Plague of Justinian 541—543 AD: a genomic analysis*. (2014). Accessed on June 22, 2014

Carter, Adam. *Black Death mysteries unlocked by McMaster scientists*. (2014). CBC Hamilton. Accessed on June 22, 2014

Recreating 1918-19 Spanish Influenza: New Influenza Pandemic Danger?

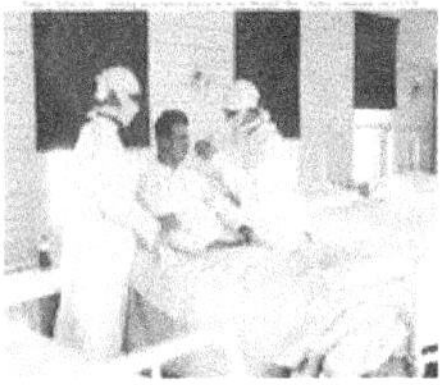

This is a Spanish influenza patient attended by masked hospital staff, US Naval Hospital New Orleans, Louisiana, USA autumn 1918.
Photograph courtesy of US Naval History and Heritage Command.

Professor Yoshihiro Kawaoka and his colleagues at the University of Wisconsin-Madison successfully reconstituted a virus 97% similar to the Spanish Flu of 1918-19, partly from fragments of avian viruses. Does this research put us all at risk?

The Great Pandemic

According to the United States Department of Health and Human Services article, *"The Great Pandemic,"* the Spanish Flu killed an estimated 40 to 100 million people in the immediate aftermath of World War One.

In response, Simon Wain-Hobson, a virologist at Paris' l'Institut Pasteur, described how this research renders the avian viruses H5N1 and H7N9, which currently circulate in Asia, more readily transmissible to mammals and argues that this research ensures that the world is now a potentially a more dangerous place.

Researchers have known the genetic sequence of Spanish Influenza H5N1 since 1997 primarily due to research carried out on the tissues of the victims of the 1918 pandemic by Jeffrey Taubenberger at the National Institute of Health, Bethesda, Maryland, USA.

The Presence of Avian Viruses Facilitated the Reconstitution of 1918 H1N1 Influenza Virus

Professor Yoshihiro Kawaoka maintains there is sufficient strain in the current avian viruses to naturally reconstitute the famous H1N1 of 1918. Professor Kawaoka found eight sequences in currently-circulating avian viruses to recreate the H1N1 influenza virus. He then tested the virulence of the reconstituted virus and found it is 1,000 times less infective than the flu of 1918.

The LD50 measure (i.e. the number of virus particles required to kill 50% of test mice) required 100,000 virus particles. Laboratory animals proved more susceptible and the virus more transmissible due to their closer, more confined living conditions. Fortunately, Oseltamvir (Tamiflu) offers robust protection against newly-created viruses.

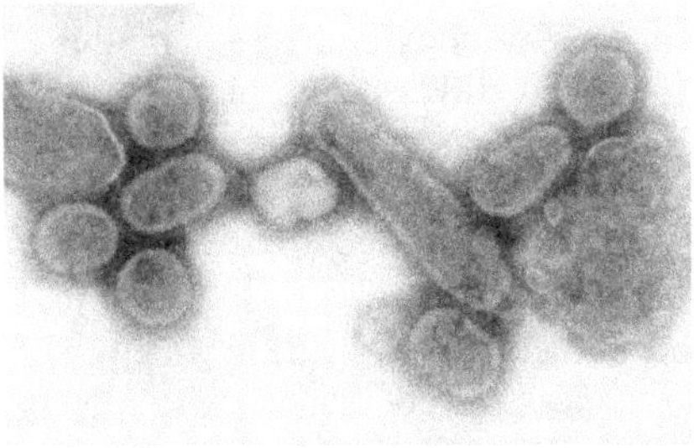

If we're hit by another Spanish Flu pandemic, health officials would prefer a vaccine that manufacturers can produce quickly. Image by the U.S. CDC

Newly-Created Viruses Could Lead to Seasonal Deaths

Simon Wain-Hobson of l'Institut Pasteur asked about '*the chances of eight viral segments reuniting in nature*", Kawaoka could not respond.

If this occurred, this combination could possibly result in 250,000 – 500,000 deaths as per seasonal influenza. In *"Do We Need to Fear Influenza, Again,"* Patrick Berche of L'Hopital Neckar in Paris asserts the level of security for these newly-created viruses was P3 and was not maximal security.

During the last two decades of research at least a dozen accidental escapes from laboratories at level P4 security have occurred.

Research into Dangerous Viruses Make No Provision for the Safety of Human Lives

Kawaoka countered Berche's warning when he insisted he used antiviral safeguards and he warned public health officials in case of a pandemic risk. Kawaoka further insisted his own vigorous security precautions in his laboratories were adequate to reduce chances of the newly-created pandemic flu virus escaping.

Fouchier's Ferrets

Kawaoka research could be less of a threat than other risky experiments. The research community feared problems after the recent revelation that Ron Fouchier rendered the Avian Virus H5N1 more transmissible between ferrets.

The creation of this 'supervirus' led to a temporary moratorium on this type of research. At the end of 2013, 56 scientific establishments combined to interdict these extreme experiments, without success.

Viruses and the Future

There are documented and verifiable occasions worldwide in which scientists work on these types of substances and compounds where there is no accountability, potentially resulting in a lethal pandemic which threatens humankind.

The scientific community must hold Professor Kawaoka, and others studying such volatile research, accountable for laboratory conditions and the safety of the public.

Resources for this article

Kawaoka, Toshihiro, et al. *Circulating Avian Influenza Viruses Closely Related to the 1918 Virus Have Pandemic Potential*. (2014). Cell Host and Microbe. Accessed on June 27, 2014

United States Department of Health and Human Services. *A History of Influenza: The Pandemic*. Accessed on June 27, 2014

Berche, Patrick. *Faut-il encore avoir peur de la grippe ? (Do we need to fear Influenza, Again?)*. (2012). Odile Jacob.

Lipsitch, Mark, et al. *Ethical alternatives to experiments with novel potential pandemic pathogens*. (May 20, 2014). National Institute of Health. Accessed on June 27, 2014

Ebola Virus Outbreak Fear in Sierra Leone, West Africa spreads to Nigeria

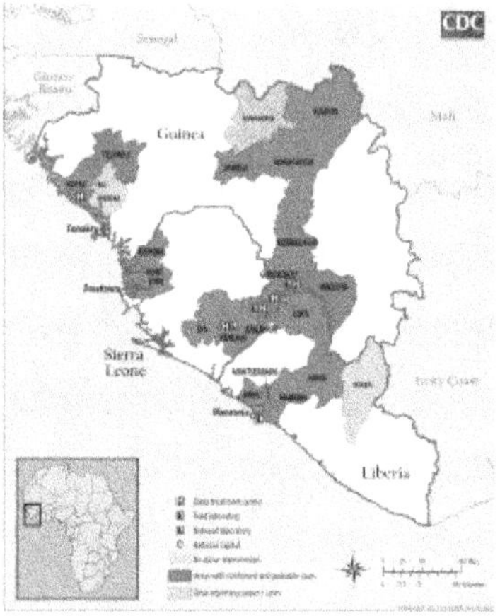

The Ebola outbreak of 2014 is spreading – will authorities be able to contain this virus? Image courtesy of the U.S. CDC.

Freetown, the capital of Sierra Leone, is in turmoil following a family's forced removal of a deadly Ebola virus patient from the rundown Harmon Road Hospital last week.

Last weekend, the patient was returned to hospital and died in the ambulance.

Ebola Update: July 30, 2014

The doctor mentioned below, who treated hundreds of Ebola patients, died yesterday.
A patient is being treated for Ebola in Germany, however Coronavirus or Marburg virus may be t he causative agent. This patient, a WHO health worker, returned from Sierra Leone last weekend.
A US WHO volunteer nurse died in Sierra Leone last week.

Ebola Outbreak around the World

Ebola, a haemorrhagic disease, has killed some 700 people in Guinea, Sierra Leone and Liberia, according to the World Health Organisation.

Sierra Leone is fighting the world's largest outbreak of Ebola virus disease. Freetown's crowded slums are a fertile breeding ground for rampant infectious diseases including cholera which killed hundreds in 2012.

A senior doctor treating Ebola patients in Sierra Leone tested positive for the virus and is being treated at a field hospital staffed by Medecins Sans Frontieres in a jungle clearing in the Kailahun district where they have reduced the normal mortality rate of 90% down to 60%. At least 12 staff members of Kenema Hospital, Freetown have died of Ebola.

Nigeria confirmed its first fatal case after a Liberian man collapsed in Lagos airport and died in hospital after testing positive.

How Do You Get Ebola?

Ebola is passed on through contact with the bodily fluids of an infected person, even after the person has died. The virus initially causes fever and vomiting and internal and external bleeding. Ebola kills 90% of those infected. There is no cure – but treatment of Ebola symptoms may improve a victim's chance of recovery.

Safety Protocols are Essential to Control Ebola

A highly-regimented safety protocol is essential to prevent the spread of Ebola in hospitals. Protective clothing for health workers is not enough to protect medical workers, visitors, or the general public.

It appears to be impossible to contain this outbreak, because the population does not respond to normal treatment protocols.

Patients do not visit doctors for suspected Ebola – most people with feverish illnesses in Sierra Leone are treated at home, which means it is difficult to determine the true extent of the outbreak.

In order to mitigate the damage, isolation wards, disease monitoring, appropriate protective equipment, improved communication, and border controls are absolutely necessary.

- Isolation wards at remote locations are essential. The remote villages of Sierra Leone, where denial, fear and rumours are aiding the spread of the disease, also prevent its victims from seeking medical help. Villagers bury infected corpses while neglecting to protect themselves. Exposed bodies pass on the virus to healthy people, thereby perpetuating the disease.
- Disease monitoring in the region is inadequate in the many scattered highland villages. The dire shortage of medical personnel means that they have little access to basic personal protective equipment, and doctors are reluctant to provide direct care for patients suspected of having Ebola.
- Access to personal protective equipment in health-care centres is vital for containing the disease.
- It is critical to implement disease control policies restricting border crossings.
- Improved communication by health officials with the media, community leaders, health professionals, and the general public is necessary to reduce misinformation and improve compliance with prevention and control measures that have been proven effective.

This 2014 Ebola Virus Outbreak is Different from the Usual Scattered, Rural, Self-Contained Disease

The scale of this outbreak is different from previous outbreaks. There have been some 700 fatalities in 2014 – a huge increase from the 20 Ebola fatalities in Uganda, 2012.

The main West African teeming urban centres of Freetown, Sierra Leone, Monrovia, Liberia, Conaky, Guinea and more recently Lagos, Nigeria are now centers where NGOs (Non Government Organisations) like Concern and MSF step in to control the outbreak before it spreads internationally. The situation is chaotic, however international agencies can help control the Ebola outbreak, and could even use mobile phones to aid surveillance of the disease.

Tempted to think that the Ebola outbreak couldn’t cause a problem in the United States and the rest of the world? International air routes link major West African cities to the U.S. – this Ebola outbreak is everyone’s problem.

Resources for this article

Ansumana, R, et al. *Sierra Leone researchers call for improved health surveillance and communication around Ebola crisis*. (2014). The Lancet. Accessed on July 28, 2014

Lethal Ebola Virus Outbreak: Your Questions Answered

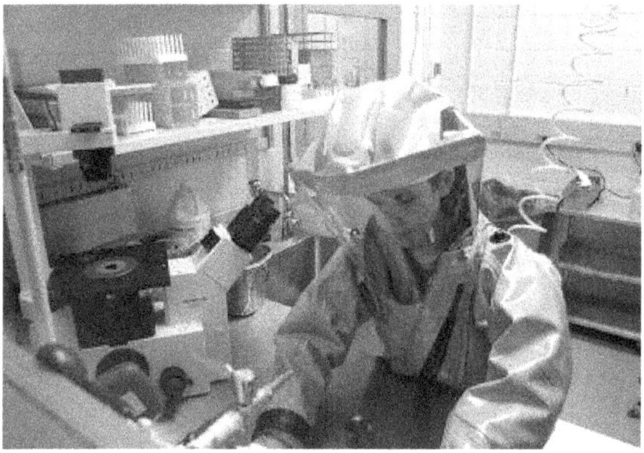

A Hazmat suited ebola worker. Do you know someone with questions about Ebola virus? Image by Decoded Science.

What does the Ebola outbreak mean for the U.S., what's different about this outbreak, and should you be worried?

2014 Ebola Outbreak: What's Different?

The usually self-contained, scattered and rural-based Ebola virus outbreak reached three West African cities (Freetown, Sierra Leone; Conakry, Guinea and Monrovia, Liberia) in early 2014 and is now festering in teeming filthy slums where hundreds died of cholera in 2013. What else is different about this Ebola outbreak?

- This is the first time Ebola has struck these regions.
- The usual locations in the Congo River basin, including Uganda in 2012, have been bypassed.
- The number of deaths in rural areas was previously less than 20. This outbreak, however, is fully and completely out of control.

Unfortunately, this Ebola outbreak retains a devastating similarity to the usual outbreaks: the 60 – 90% fatality rate. Health officials use isolation tented areas away from cities to deal with those infected with Ebola virus.

Containing Ebola: Relatives, Health Care Workers

When someone has any contact with an infected victim, even after death, that person must be monitored for the virus. This creates a huge pool of potentially-infected people to track.

For example: If the 700 Ebola virus deceased contacted 100 relatives and health workers each, which is quite feasible, then 70,000 people must be isolated.

In Sierra Leone, Dr. Khan died of Ebola after contacting at least 30,000 patients. That means health officials must monitor and isolate 100,000 individuals – an impossible task.

There's a 21 day incubation period, during which an apparently healthy individual can end up bleeding profusely from every orifice. The concept is tragic and difficult for health workers to deal with, yet, WHO international medical practice requires those infected -and those who were in close contact with them- to be kept in isolation until the incubation period elapses.

This is impossible. And so the virus spreads.

The WHO described the response by Guinea, Liberia and Sierra Leone as wholly inadequate, and declared the Ebola outbreak in West Africa as a Global Public Health Emergency, August 8, 2014. This action will shut down at least three West African countries, in hopes of isolating the outbreak.

Ebola Outbreak: Quarantine

The situation is dire in Sierra Leone with at least 932 dead from Ebola virus as of August 7, 2014. Dealing with returning aid workers requires quarantine back in the United States.

A US doctor who volunteered to work with MSF in Guinea for two weeks in May/June 2014 described dealing with Ebola sufferers whilst clad in a plastic mesh suit with goggles and a hood in 35 degree Celsius heat and 90% humidity – and witnessing a mother and son who died within 24 hours of each other without the benefit of a human touch.

Such US doctors, or European medical personnel, returning from ebola risk areas in West Africa should be quarantined.

Currently two US medical professional who contracted Ebola virus whilst working with Ebola victims in West Africa have been transferred to Atlanta, Georgia, USA where both are being treated with monoclonal antibodies, and fortunately are responding to treatment.

Aid workers struggle are struggling to contain the Ebola outbreak in Guinea, West Africa. Image courtesy of the U.S. CDC.

Ebola Questions and Answers

The Ebola outbreak has many people asking questions. Let's provide some answers:

How do healthcare workers catch Ebola, since they're wearing protective gear? Doctors and nurses succumb to Ebola, despite full body cover, when they scratch an itch by moving their goggles or by exposing their forearms, and so on.

How is Ebola transmitted? Can you catch Ebola if someone sneezes on you? Ebola is not transmitted through sneezing- an aerosol – but only through touching the body fluids of an infected person.

Are there any treatments for Ebola other than supportive care?

Currently, 2 US health workers with Ebola are responding to monoclonal antibody treatment in Atlanta, Georgia, USA. Monoclonal antibodies are lab-produced, and engineered to attack a specific type of cell.

Is Ebola contained, or has it spread beyond Africa?

There are 4 known Ebola sufferers located outside Africa. One Ebola victim, contracted in Sierra Leone, died in Saudi Arabia last weekend, marking the first death from Ebola outside Africa. In addition, a Spanish priest, an Ebola victim, was transferred from Sierra Leone to Madrid Spain last weekend.

What Kind of Virus is Ebola?

The Ebola virus is a filovirus; it looks like a piece of string, or filament. Ebola originates from the ubiquitous African fruit bat and is a zoonosis – that's an animal virus that can transfer to humans. Ebola is self-limiting – not considered a 'successful' virus, because it kills the host. Eventually, Ebola will run out of victims.

Will Ebola mutate to become more successful, and decimate the population?

Ebola does not mutate.

Could Ebola be a Biological Weapon?

The United States retains Ebola virus at Level P4 Laboratories where only the most trusted virologists are allowed access to the virus. Russia has a vaccine from the antibodies generated by survivors of Ebola virus. These two facts confirm Ebola virus could be used as a bio-weapon.

What Will Ebola Mean for the United States and Europe?

The U.S. and U.K. public health systems maintain a state of readiness to prepare for potential outbreaks of high-priority biological agents rarely seen – including Category A diseases like Ebola, Smallpox, and the Black Plague – before an outbreak ever actually arises.

High-priority agents include organisms considered a risk to national security. How could illness become a national security risk? These are the diseases that can be transmitted from person to person and result in high mortality rates – this set of circumstances could potentially cause public panic and require special action for public health preparedness.

Here's the bottom line: The robust US and EU health systems can deal with Ebola. Isolation of the Ebola-struck areas, and quarantine of everyone who comes in contact with an Ebola victim will ensure that the outbreak doesn't extend to other nations.

Resources for this article

U.S. CDC. *Category List of Category A Diseases*. (2014). Accessed on August 08, 2014

Karimi, F, et al. *WHO: Ebola outbreak in West Africa an international health emergency*. (2014). CNN. Accessed on August 08, 2014

Qiu, X, et al. *Successful treatment of ebola virus-infected cynomolgus macaques with monoclonal antibodies*. (2012). Science Translational Medicine. Accessed on August 08, 2014

Ebola Virus: a Return to its Origins

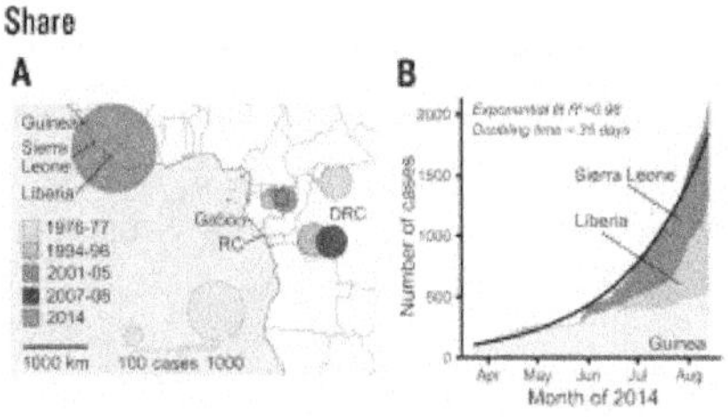

Ebola incidence 1976-2014 Graph A. Ebola doubling time (35 days). Image from Science August 2014

1,900 Ebola victims have died since the outbreak commenced in February 2014, in Guinea, West Africa.

At least 700 people have died in Sierra Leone, Liberia, Guinea, Nigeria and the Democratic Republic of Congo within the last three weeks, confirming the Ebola outbreak is fully and completely out of control.

Air travel suspended to and from West Africa prevents NGOs from sending aid such as hygiene suits to Sierra Leone.

Riots stopped an attempt to quarantine a district of Freetown, Sierra Leone. 230 deaths weekly could result in the 15 weeks remaining in 2014 multiplied by 230 deaths = 3450 probable fatalities. Of course, we can’t know the exact number of people who are infected in 6 West African countries.

A total of 20,000 deaths is probable and quite possible from this Ebola outbreak within the next year.

Origins of the Ebola Virus

The spread of Ebola in West Africa is caused by at least five different species of Ebola virus.

A co-operative investigation by the Broad Institute of Cambridge, Massachusetts, USA and Harvard University, Boston, USA on the genome of Ebola incorporating research on the Ebola victims in Kenema Government Hospital, Sierra Leone, determined that the Ebola virus originated in Guinea and spread rapidly to Sierra Leone.

The research was conducted by Stephen Gire and Augustine Goba, who were in Sierra Leone and witnessed the first death in Kenema Hospital, at the end of May, 2014, and were able to diagnose the victim.

A visit to neighbouring Guinea allowed a further 30 infected women to be examined and traced to their burial site. 66 more patients in Kenema hospital, of whom 70% died by mid June 2014, were traced and investigated in Sierra Leone.

The researchers examined and compared virus samples from these victims were compared with the existing data from the former epidemics and the genesis of the original outbreak.

The Ebola epidemic samples from 1976 were available from the Central African region – researchers deduced that a new variant arose from 2004 which then forged ahead during the last decade to finally generate the current February 2014, Guinea Ebola outbreak.

The African Fruit Bat Harbours the Ebola Virus

The transfer of Ebola virus from the natural reservoir of the African Fruit Bat assumes they are consumed in a poorly cooked state. Bush meat is the staple diet of many West Africans, who are casual in handling the meat. The researchers confirmed two people handling the unprotected body of an Ebola victim at their burial ensured at least one of the handlers became infected. This is a daily occupational hazard.

New Ebola Virus Transmits Easily

This Ebola outbreak is distinguished by the detectable and verifiable ease with which the Ebola virus is transmitted. The readily identifiable change in the virus, whereby a new species of Ebola arose within the last decade is responsible for this current rapid infection with numerous victims – unlike the 2012 Ugandan outbreak which killed less than 20 people.

Researchers have mapped the current Ebola virus in order to provide essential data for the production of the experimental ZMapp, a mixture of three antibodies to Ebola, which is manufactured by the Californian BioTech Firm, Mapp Biopharmaceutcal. The company provided three Ebola victims, two US and one UK health worker, with the antibodies, which gave them a chance to fully recover from the Ebola virus.

Ebola Outbreak Future

Western volunteer health workers should be placed in quarantine for at least 21 days, on their return. This alarming, out of control, outbreak is one that West African hospitals cannot deal with.

Resources for this article:

Gire, S, et al. *Genomic surveillance elucidates Ebola virus origin and transmission during the 2014 outbreak*. (2014). Science. Accessed on September 05, 2014

Ebola virus victims could total 1,000,000 within 12-18 weeks

Liberian President Ellen Johnson Sirleaf appeals for more international aid to combat ebola in her ravaged country. Image courtesy of the U.S. White House

Ebola is a rare hemorrhagic disease that is currently out of control in West Africa. The total number of outbreaks to date claimed 1700 lives, from 1976-2013.

The current outbreak, with 2,246 fatalities by August 31, 2014, exceeds all the previous outbreaks put together. What can the global community do to halt the spread of the disease? If more aid does not arrive, the victim count could reach 1,000,000 victims in the next 12 to 18 weeks.

Ebola Virus History

The first outbreak was located beside the river Ebola, in the Democratic Republic of Congo (DRC) in 1976. Dr. Petr Piot of Antwerp, Belgium, received Ebola blood samples in ice from Ebola victims in the DRC by post, in 1976, prompting him to visit some Belgian nuns in charge of a hospital in the DRC where there were no doctors.

In this hospital, caregivers were washing and re-using syringes because they had no health supplies. Dr. Piot named the virus from the river.

The realisation that Ebola could be an effective biological weapon ensured that health officials from various nations sought and secured a vaccine. The US, Canada and Russia were engaged in harvesting antibodies from Ebola survivors, so that an effective vaccine against this global threat could be created.

Dr. Kobinger, in Ontario, Canada, developed a vaccine for the rare Ebola virus using antibodies from survivors of the 1976 Ebola outbreak. Perversely, any alteration of the genome may have generated a more virulent Ebola subspecies! These countries harboured Ebola, in secure locations, along with the related Marburg virus, so that they could produce vaccines quickly if necessary.

Ebola Outbreak: Hemorrhagic Infection Spreads

Ebola uses protein spikes to attach to red blood corpuscles. Ebola then enters and kills the cells – producing a hemorrhagic flow from all organs. This flow of blood is highly contagious to another person touching it. Antibodies, when introduced into a person infected with Ebola, retracts these proteins so that the Ebola virus cannot attach to the cells.

In Monrovia Liberia, Ebola victims die at the gates because there are no more beds. In West Africa there is one doctor for every 71,000 patients. The U.S. CDC in Atlanta Georgia is monitoring the Ebola outbreak, and recently sent a team to Senegal, West Africa.

Could U.S. involvement spur a global movement to assist in West Africa? The reproduction of ZMAPP requires tobacco plants to reproduce enough implanted Ebola antibodies/proteins to generate enough antibodies to generate enough vaccine to control the Ebola outbreak. The next batch to treat the West African outbreak will be ready in December 2014. What will the Ebola outbreak look like by then?

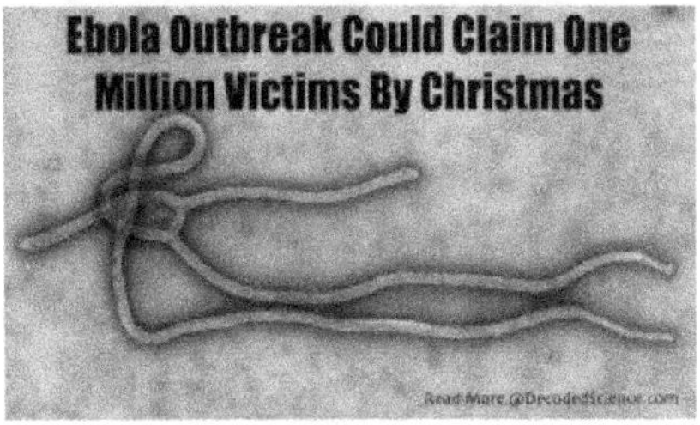

The global community must come to the aid of West Africa to stem the spread of Ebola. Image by Decoded Science

The 2014 Ebola Outbreak is fully out of Control

Each Ebola victim generates at least one other Ebola victim. The doubling time for Ebola infections is now down to 4 weeks.

Within one month the doubling time will be 7 days. The current number of Ebola victims – over 2200 – will double to 4,000 in one month, then to 8,000 and then 16,000 as we reach a 7-day doubling time.

Taking into consideration the addition of more beds, more medical assistance, and the availability of ZMapp to the West African Ebola victims, the doubling time may not increase as quickly. With an outpouring of aid from the global community, the Ebola outbreak may require 8 weeks to reach a seven-day doubling time.

When we do reach the 7 day critical doubling time, however, it's a short interval to reach a million victims – this could happen within 12 to 18 weeks from now.

What's next for Ebola?

Clearly, integrated international aid, more international volunteers, more hospital beds, and expanded use of treatments such as ZMapp could make a world of difference. What if the aid doesn't appear? The victim count will continue to increase.

October 15, 2014 at 6:16 am

Proceedings from Dallas USA convention on Ebola:- All travel to and from West Africa should be suspended, the ebola infected areas isolated and a vaccine introduced for all western medical and military personnel operating in the area. The Ebola vaccine will be available with 4 months. Ebola, a hemorrhagic disease, destroys the liver which allows blood to clot, the end result is a shedding of 5 litres of blood exiting the stricken patient daily. Ebola a category 4 pathogen requires just 2 viral particles to induce an Ebola infection. 10 billion Ebola particles are present in 1 ml of blood rendering Ebola highly infectious. This Ebola outbreak is costing West African countries $30 billion annually. It is expected the Ebola virus will become less virulent because killing the host does not allow the Ebola virus to survive and thrive. Airport screening will not detect healthy non-overt Ebola sufferers.

Reply

October 24, 2014 at 3:47 am

Sierra Leone, with a population of 6 million, had 160 doctors dealing with the 2014 Ebola outbreak. 60 of these doctors died from Ebola while treating the outbreak. At present the time lag between the massive western response to Ebola in Sierra Leone ensures there is a significant time lag between deaths from Ebola and those infected. There remains an unknown number of Ebola sufferers in West Africa. A structured US and EU military co-ordination with isolation and separate treatment facilities in West Africa will inevitably contain the Ebola outbreak by mid-December 2014.

Nigeria with 20 Ebola cases and 8 deaths has had no further Ebola outbreaks so that the WHO declared Nigeria Ebola free.

US and EU medical personnel returning from West Africa must be screened and monitored for Ebola.

Reply

Emily says

October 8, 2014 at 10:20 pm

Where did you get this statistic? "In West Africa there is one doctor for every 71,000 patients."

Thank you.

Reply

October 10, 2014 at 3:50 am

UN, WHO, UNESCO, UNHCR all confirm statistics regarding medical; personnel per head of population.

Reply

1. Jordi says

   October 7, 2014 at 1:13 pm

   Congratulations on your article, Professor … I had reported the disease in the media and Internet, but now recently became aware of the global dimension of the problem … The report is clear and very revealing … and sorry if I made a gross error in my previous comments … .Jordi

   Reply

2. Jordi says

   October 7, 2014 at 12:56 pm

   Congratulaciones por su artículo, Profesor …Me había informado de la enfermedad por los medios y por Internet, pero ahora recién tomé conciencia de la dimensión mundial del problema…Su informe es claro y muy revelador…Y disculpe si he cometido algún error grosero en mis comentarios anteriores….Jordi

   Reply

3. Darla says

   October 6, 2014 at 11:59 am

   Is there an update on this situation? I would appreciate hearing more if you are still tracking the spread of this virus.

   Reply

October 7, 2014 at 4:38 am

Update pending October 7, 2014

Reply

October 9, 2014 at 4:46 am

Between February and September 2014, 3,450 died from Ebola. During the same time period 300,000 died from malaria and 800,000 died from tuberculosis. Should 10 people contract Ebola then 3 will survive this is nature of a virulent Hemorrhagic Fever Virus (HFV).
Ebola will be contained with US and UK troops building hospitals, isolation & treatment centres in West Africa which will interdict the rise in Ebola within 6 to 8 weeks. No one can predict the future, however, the world has woken up to Ebola and the civil war wracked Sierra Leone and Liberia infrastructures need international help.
Nigeria has dealt effectively with Ebola through isolation and re-hydration treatment of Ebola victims, similarly, Senegal has dealt effectively with one Ebola case.
A nurse in Madrid, Spain, double taped her glove which she released with her uncovered hand.
The US must screen West Africans or travellers to and from the US.

Reply

4. Jordi says

   October 4, 2014 at 6:00 pm

   Ebola is a "primitive virus" in humans, still not evolved to "not kill its host" … When it becomes chronic in humans lower your degree of lethality, like other virus … It would sense the evolution of a virus that kills all guests ?… The Junin virus (or "mal de los rastrojos"), if it comes in the cities, could also evolve into a variant contagious among humans … Luckily we have the vaccine from the 50 …

   Reply

5. Jordi says

   October 4, 2014 at 5:54 pm

In a way, we are better prepared than the first world countries, they never had before hemorrhagic diseases … In Argentina there is experience in treating patients with Junin virus, an hemorrhagic virus family also has Ebola … High mortality rates, but a vaccine is now available … But with this disease, Ebola, one can only fight the symptoms as they arise … viruses do not have remedies in the style of an antibiotic is only … using antivirals to contain the attack, expect the patient to support the disease until the virus complete its infectious cycle, and pray that an effective vaccine is … The high mortality is mainly due to the poor state of health in Africa and systemic deterioration of patients when they come to the hospital … The Junin Virus (or "mal de los rastrojos") have lower mortality rates, and perhaps more manageable through the knowledge of our doctors, who are the real heroes movie … I just hope they serve this previous experience, when the inexorable arrival of Ebola River Plate to happen, and it will in a similar way to America … God help us to argentos and americans too…

Reply

6. Jordi says

October 4, 2014 at 5:44 pm

En cierta forma, estamos mejor preparados que los países del primer mundo, que nunca antes tuvieron enfermedades hemorrágicas…En Argentina hay experiencia en tratar pacientes con el Virus Junín, que es un virus hemorrágico de la familia del Ébola…También tiene altas tasas de mortalidad, pero ya hay disponible una vacuna…Con esta enfermedad, el Ébola, sólo queda luchar contra los síntomas a medida que estos aparecen… Los virus no tienen remedios al estilo de un antibiótico…solo queda usar antivirales para contener el ataque, esperar que el paciente soporte la enfermedad hasta que el virus termine su ciclo infeccioso, y rogar que se encuentre una vacuna eficaz…Las altas tasas de mortalidad se deben mayormente al estado deficiente de la sanidad en África, y al deterioro sistémico de los pacientes cuando acuden al Hospital…El Virus Junín (o mal de los rastrojos) tiene tasas menores de mortalidad, y tal vez sea mas manejable, por el conocimiento de nuestros médicos, que son los verdaderos héroes de la película… Sólo espero que sirva esta experiencia previa, cuando ocurra el inexorable arribo del Ébola al Río de la Plata, y lo hará en forma parecida a EE.UU…Dios nos proteja a los argentinos…

Reply

7. David says

September 28, 2014 at 9:17 pm

The article states that the doubling time is getting shorter. Why?

Reply

October 1, 2014 at 4:18 am

Each Ebola victim generates another Ebola victim due to personal contacts. The number of victims reached 10,000 – both infected and dead at the end of September 2014. The 21 day incubation period (three weeks at present) determines the doubling time. In three weeks time the 10,000 will generate 20,000 Ebola sufferers.

The Ebola outbreak is fully and completely out of control in Sierra Leone.

Many West African refugees arriving in Southern Italy may be Ebola sufferers.

Reply

- Jay says

  October 2, 2014 at 10:43 pm

  Doubling time does not change based on the outbteak size. One infects 2, 2 to 4, 4 to 8, 8 to 16 etc, but each time its been taking the exact same time to double. What you are doing is doubling the outbreak each time AND halving the time it doubles again. Thats not how it works. The outbreak spreads twice as fast, but the doubling time remains unchanged. If u plot the current outbreak since march, the cases double almost exactly every 26 days.

Reply

October 4, 2014 at 4:40 am

The USA is listed among those countries with Ebola victims due to the Liberian man who presented symptomless this week and was turned away only to return to the same Texas, USA hospital who then belatedly treated him. Unfortunately, this man was in contact with at least 100 other people. It is a priority to trace these contacts.

Those Ebola victims within the 21 day incubation period for Ebola could consist of those victims within 7, 14 and 21 days respectively of displaying symptoms of infection. The three week (21 day) current doubling time could be eclipsed by these three (7, 14 &21) incubation times overwhelming treatment centres.

Ebola at this exponential scale simply cannot be comprehended, this is precisely why the epidemic is fully and completely out of control in Sierra Leone.

Reply

8. says

   September 17, 2014 at 5:01 pm

   Its means he is not totaly cured yet..
   I dont consider there is a cheap cure..other wise its a Qurban for the Dajjal.

   Reply

9. Dr. Jack Butler says

   September 16, 2014 at 10:03 am

   I have a 98% successful treatment for Ebola. I am willing to go to West Africa and administer this treatment MYSELF. However I lack funding and also lack authorization from a local government that will allow me to give this treatment to local patients. It will take about $20,000 USD to treat and cure 200 patients. I propose that this be done as a trial. The results will then be published if it is successful.

   Dr. Jack Butler

   Reply

   - David says

     September 16, 2014 at 1:45 pm

     How could you possibly have a 98% effective treatment for Ebola? In what population did you test this "treatment" to arrive at your 98% rate?

The World Looks to the USA for Leadership in Countering the Ebola Outbreak

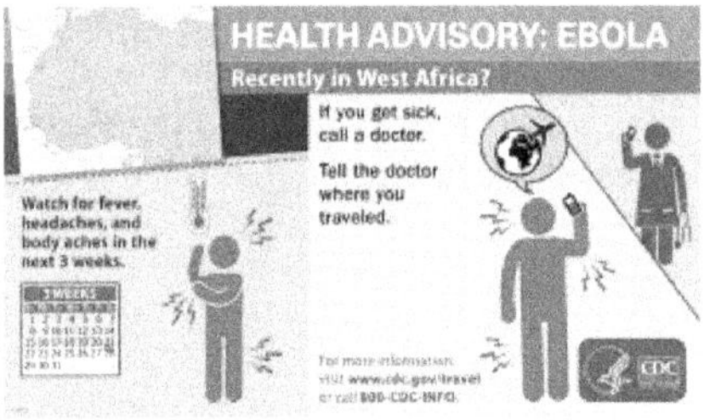

The Ebola outbreak means that travel could be more dangerous. Image courtesy of the U.S. CDC

'*U.S. Leadership is the only constant in an uncertain world*,' according to U.S. President Barack Obama. Does that also apply to the Ebola outbreak in West Africa?

Ebola Outbreak Continues

The scourge of Ebola continues in West Africa, with 1,199 cases in Guinea – and 739 dead. Nigeria has 20 cases and 8 dead.

Liberia has 3,834 cases and 2069 dead, and Sierra Leone has 2427 cases and 623 dead- these numbers confirm that Ebola is out of control. Liberia and Sierra Leone have been through difficult civil wars and their infrastructure cannot cope with normal cases of TB and typhoid which kill large numbers annually – much less this Ebola outbreak. Senegal now has 1 case and no deaths.

In the U.S. – a man who travelled to Nigeria presented with symptoms of ebola in Washington, and hours later a patient presented with '*flu-like symptoms and a travel history matching criteria for possible Ebola*' in Maryland. And a Liberian man is in isolation in Texas with Ebola.

Texas Health Presbyterian Hospital September 25, 2014

A Liberian driver, who did not inform Texas Health medical staff that he had been in contact with Ebola, told a nurse he had travelled to the United States from Liberia when seeking medical care for Ebola symptoms. Unfortunately, he was not diagnosed with Ebola at that time, and was instead sent home with a course of antibiotics.

This, the first case of Ebola in the U.S. – and the inadequate response by health officials – have increased fears the disease could spread throughout the country.

The Ebola victim eventually returned to the hospital on September 28, suffering from diarrhoea and vomiting and was re-admitted. He was then put into isolation. At least four people he'd been in contact with are now under quarantine – with a guard making sure they stay home after they repeatedly broke quarantine.

Can the U.S. Medical System Handle Ebola?

'The U.S. has the most capable health care infrastructure and the best health care workers in the world'

This quote is attributed to Ms. Lisa Monaco, President Obama's very confident counter-terrorism adviser.

On the other hand, U.S. workers are building hospitals, laboratories and treatment centres in West Africa. Many U.S. health workers have direct experience of treating Ebola, having volunteered to work for Medecins san Frontieres (MSF) for 4 weeks at West African treatment centres. Unfortunately, many returned with minimal screening – even though they worked with Ebola patients for at least one month, putting them at risk of infection. If Ebola enters the U.S. en masse, the medical system will be tested.

Ebola, West Africa, and the Outbreak: The Future

West African drivers, farm workers, taxi drivers etc., have succumbed to Ebola, so fields are not tended and supplies are not being delivered by truck. To add to the problem, food is now in short supply in West Africa.

This writer has testimony from Irish Concern and MSF workers that the Sierra Leone and Liberian economies are collapsing. Outside intervention must be robust to succeed.

Ebola Doubling Times:

The same error regarding the doubling time recurs. The number of ebola victims in the initial outbreak remains unknown, so that 2,4,...512, 1024 doubling cannot apply. The 7, 14 & 21 weekly incubation times could find a large amount of Ebola victims coalescing so that no treatment centre could cope. The world has woken up to Ebola... But, thanks to the available health systems, help from other nations, and medical technology, this writer believes this epidemic will be contained.

U.S. Aid Reduced Ebola Deaths in Liberia by 50%

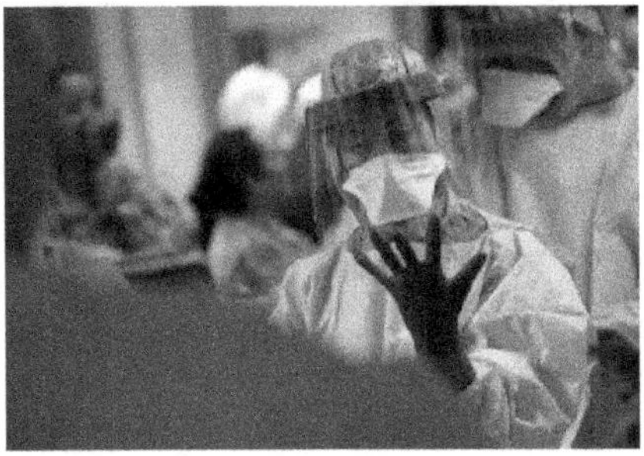

The U.S. Military interventions in the Ebola outbreak may have turned the tide in this year's unprecedented Ebola outbreak. Image by Master Sgt Jeffrey Allen

Could interventions have turned the tide of the Ebola outbreak?

In Liberia, by October 23, the total of 2,705 dead from 4,665 confirmed Ebola cases is more than half of the total of registered deceased from Ebola (4,992) and so seems to indicate a decrease.

The total death toll should, in fact be 2 x 2705 i.e. 5410 – due to the normal 4 week doubling effect.

The Liberian Red Cross, charged with collecting the corpses around the capital, Monrovia, stated that on October 28 the number of deceased, which reached a peak of 300 per week by mid-September, declined to 117 just last week.

The Secretary General of the Liberian Red Cross, Fayah Tomba, warned *'not to declare a victory because the number of cases continue to be under-declared, particularly in Monrovia'* (translated from an article in Le Monde.Fr newspaper).

International Co-operation

The United States intervention concentrates on Liberia with an Infantry Brigade of 4,000 soldiers building isolation and treatment centres in forested areas of the country. The mission has seen success, as death rates appear to be going down.

The United Kingdom is concentrating on its former colony of Sierra Leone with a 750 strong battalion and a Royal Navy supply ship in support, again with some success in declining numbers of victims.

French government concerns in Guinea – namely through health, diplomacy, security and research directed by the President of the Pharmaceutical Company Aviesan along with Medecins sans Frontieres (MSF) – intend that by mid-November, a clinic administering an experimental antiviral favipiaravir will be available by January 2015 to treat Ebola sufferers.

France will concentrate on facilities in Guinea in co-operation with NGOs such as Goal, Concern and MSF. The main centre will be located close to Conakry with French army and Guinean army assistance, caring for between 30 and 200 patients.

The present Red Cross centre at Macenta, Guinea located near Liberia and Sierra Leone, with 100 beds, has access and security problems.

At the end of December 2014 or early 2015 there will be several 30 – 50 bed facilities for the treatment of ebola in forested areas, sufficiently secure to be effective.

Ebola Treatment and Prevention

Favipiaravir (T-705) represents the first effective therapeutic agent for advanced Zaire EBOV (Ebola virus) infection in an animal model. It reduces viraemia (reduces the number of viruses), ameliorates clinical and biochemical signs of disease, and prevents lethal outcome in 100% of the animals if treatment is commenced 6 days after infection – that's 2–4 days before the time of death in control animals. Human clinical trials will follow.

Currently a large scale clinical trial of an Ebola vaccine on 400 volunteers is underway in Switzerland.

Ebola Quarantine Measures: International Opposition

International agencies oppose systematic quarantine measures for medical personnel who had contact with Ebola victims, simply because this discourages medical personnel from travelling to West Africa where they are so desperately needed.

In reality, there is no possibility of an Ebola epidemic outside West Africa due to differences in medical and sanitary standards, and aid from outside nations, so quarantine is unnecessary.

A large French civilian population in Cote d'Ivoire, for example, ensures France will aid their compatriots should an outbreak occur as both Guinea and Liberia share a border with Cote d'Ivoire.

Ebola Outbreak: U.S. Involvement Has Decreased Deaths

From mid-September 2014, the now six-week U.S. building of isolation and treatment centres has resulted in a decrease in deaths from Ebola. Ebola can be contained – however it will not be eliminated. Uganda functions with Ebola infections monthly and is able to contain the outbreaks in rural areas. Nigeria is declared Ebola free by WHO, and Mali and Senegal have successfully contained one case each of Ebola. Ms. Samantha Power, the Irish-born US Ambassador to the UN toured the Ebola affected countries in West Africa last week and expressed her satisfaction with their progress in containing Ebola to date. In general, the outlook is improving.

Resources for this article

Benkimoun, Paul. *Ebola, Il n'y a pas de mobilisation europieenne*. (Octobre 30, 2014). Le Monde Fr. Accessed on November 05, 2014

Watson, Paul Joseph. *Christian Patriots/ Ebola (Hazmat) Suits*. (October 2014). Christian Patriots. Accessed on November 05, 2014

Oestereich, Lisa, et al. *Favipiaravir T705*. (February 2014). Science Direct . Accessed on November 05, 2014

Dosso, Zoom . *Liberian Red Cross*. (October 28, 2014). Yahoo News. Accessed on November 05, 2014

November 15, 2014 at 4:56 am

In striving to accentuate the positive, US, UK and French aid to West Africa was mentioned in this article. The robust US aid in Liberia has had results within 6 weeks of building hospitals in the bush. UK aid in Sierra Leone, a country with tremendous mineral wealth and tourism potential, has had education and sport (contacts so minimised) halted since July 2014 and recorded an increase in Ebola deaths in September 2014.

Reply

Richard says

November 14, 2014 at 2:50 am

Yes, as usual, only aid from the U.S. is mentioned.

Bird Flu and Ebola Viruses: Crossing the Species Barrier Should Not Be Ignored

This Mute Swan could harbour Bird Flu. Image by AcrylicArtist

Bird flu and Ebola are animal viruses with a proven ability to infect humans and kill them: They are zoonoses.

This week, scientists identified the strain of bird flu causing the North of England (Yorkshire), Nederlands and Germany outbreak as H5N8, which is related to the H5N1 bird flu virus and requires immediate action.

The H5N1, 2003 bird flu virus, is lethal, killing 60% of humans infected.

Bird Flu vs. Ebola: Zoonoses

Zoonoses are animal viruses which did not evolve to live in human hosts; but can infect humans. Zoonoses cause the most deadly infections known to humankind. Examples include rabies, Ebola, SARS and bird flu.

Both Ebola and bird flu replicate to overwhelm the human immune system which results in alarming mortality. The human immune system cannot deal with zoonoses.

- Bird flu is difficult to contract, requiring direct contact with poultry, and has killed 400 people since 2003. Bird Flu viruses take advantage of highly intensive poultry farming on which humankind depends. Bird flu is readily spread by wild birds and has been quiescent since 2003. The current bird flu outbreak in Northern England and the Nederlands does not threaten humans, but we should not ignore the bird flu.
- Ebola, equally difficult to contract, and as shown with previous outbreaks is significantly rare. Ebola is now proving lethal in West Africa. Scientists first identified Ebola, an animal virus , in 1976. They showed that most outbreaks were linked to bush-meat consumption in Central Africa. Population pressure on the jungles of southern Guinea where the 2014 Ebola outbreak started are extreme. New roads in virgin jungle clearances allowed Ebola to move from the jungle to West African city slums.

Ebola in West Africa, November 2014 Update

Medecins Sans Frontieres (MSF) and the World Health Organisation (WHO) have tentatively identified Patient Zero, a 2 year-old boy who died in Meliandou, Guinea, West Africa. This identification makes it possible to identify the location, distribution and finally confirm the species of Ebola causing the current outbreak.

The massive US aid for Liberia resulted in a reduction of Ebola deaths by November 2014, but in Sierra Leone, a lack of co-operation from the government thwarts the British effort to combat Ebola. The government did not release UK aid delivered to the airport in August 2014 until October 2014. Currently there are not enough beds for Ebola victims in Sierra Leone.

- WHO declared Nigeria, with 20 cases and 8 Ebola deaths, Ebola-free in October 2013.

- Senegal, with one Ebola case and no deaths, fully contained Ebola.
- An Imam in a mosque in Mali died of Ebola; mourners touched the body and 4 contracted Ebola.

Why was this Ebola Outbreak out of Control?

The MSF warned WHO in April 2014 that the Ebola outbreak was out of control. The WHO dismissed this claim. In August, MSF stated that none of the organisations in the most affected countries [the UN, WHO, local governments, NGOs (Non Governmental Organisations, including MSF)] could deal with the serious outbreak of Ebola. The massive US response in Liberia, the struggling UK aid to Sierra Leone, and French aid to Guinea have finally taken effect.

Conclusion: WHO now Fully in Control, November 2014

By October 27 there were 4,922 deaths from Ebola with 13,703 cases. In WHO's Geneva, Switzerland, headquarters, October 28, 60 senior WHO officials track and monitor Ebola cases as they deploy more doctors, nurses and experts to the Ebola zone. This Ebola outbreak will be contained with WHO co-operation by January 2014.

The WHO have reacted in a timely manner and are aware of the potential threat in the European bird flu outbreak as well.

Resources for this article:

Clarke, Tom, Science Ed.. *As we focus on Ebola, we underestimate bird flu at our peril - See more at: http://blogs.channel4.com/tom-clarke-on-science/focus-ebola-underestimate-bird-flu/1495#sthash.3BRBVYIH.dpuf*. (2014). Channel 4. Accessed on November 21, 2014

Cheng, Cheri. *Counsel & Heal*. (2014). Counsel & Heal. Accessed on November 21, 2014

Walt, Vivienne. *Missing in Action*. (November 14, 2014). Time Magazine, Vol 184.

Ebola Virus Disease Democratic Republic of Congo and Uganda, August 2018 - March 2020

The Democratic Republic of the Congo is grappling with the world's second largest Ebola epidemic on record, with more than 2200 lives lost and 3400 confirmed infections since the outbreak was declared on 1 August 2018. The outbreak is occurring in North Kivu and Ituri provinces. Neighbouring countries are taking steps to mitigate the risk of spread. The World Health Organization has hundreds of staff on the ground supporting the Government-led response together with national and international partners. Since there is little reporting of this EVD outbreak, then, this WHO report is required.

Latest numbers as of 30 March 2020

3453

Total cases

2264

Total deaths

1169

Survivors

Total of **3453 cases** (3310 confirmed & 143 probable), including **2264 deaths**, **1169 survivors**, and patients still under care.

Source: Ministry of Health, Democratic Republic of the Congo

Novel Corona Virus, Covid-19, 2020

This New Corona Virus originated in Wuhan city c. 31 December 2019. The virus was isolated, identified as SARS-Cov-2 and is related to SARS of 2002 but less contagious and notified worldwide unlike the SARS (Severe Acute Respiratory Syndrome) outbreak in 2002. The number of deaths from Covid-19 at 5% is less than SARS of 2003 at 9.6%. Deaths may reach 50,000 early April, 2020 with 1,000,000 cases in 185 countries ensures a mortality rate 5.0%. South Korea, Iran and Italy are severely affected. Covid-19 is in every continent except Antarctica. A more refined test for Covid-19 resulted in a sharp increase in deaths. Western countries evacuated their citizens from Wuhan. Treatment primarily depends on isolating the infected patient which required China to build 2x 1,000 bed hospitals within two weeks. New Corona Virus may have originated c. December 8 2019 according to a Chinese Doctor, since deceased.

A worldwide medical emergency for Covid-19 was declared January 31, 2020. Covid-19 is present in all Chinese provinces with probably millions infected and person to person transmission a factor. The elderly and those with underlying conditions are at risk from Covid-19. Irresponsible people spread this infection with a non-caring attitude. China is managing Covid-19 quite well in late March 2020 or so we are led to believe.

Covid-19 causes viral pneumonia, those infected develop a cough, fever and breathing difficulties, essentially viral pneumonia. Goblet cells and cilia cells in the lungs are targeted by the virus and cannot clear the airways resulting in congested lungs and death. Hospital admissions secure breathing support and fluids. Recovery depends on the strength of the patient's immune system. The Chinese New Year, the year of the Rat, ensured airports were a focus for the spread of the virus worldwide. Globalisation is the cause of Covid-19 spreading.

The total amount of people in China infected with Covid-19 is unknown because hundreds of people with pneumonia symptoms were turned away from hospitals due to a lack of hospital beds. The wildlife market where the first victims were infected is located beside the high speed railway station where trains leave for Hong Kong and cities throughout China. A travel ban and lock down of populations were imposed and arrested the spread of Covid-19. The transmission of Covid-19 has a person to person element with carriers without symptoms infecting millions of people.

This is an entirely new respiratory virus about which very little is known. The doubling effect due to the 12 day incubation period has become apparent. The panic and hysteria is difficult to deal with. The Tokyo Olympics are deferred until 2021. A pandemic describes Covid-19 appearing in many countries without direct contact.

The daily doubling of Covid-19 in Ireland by early April 2020 is due to the 12 day incubation period of the virus and is similar to the Italian situation. 61 million people in Italy are in lock-down. The distancing factor on the island of Ireland is crucial to contain Covid-19. A interferon type treatment from SARS may interdict viral transmission.

Donny from Queens banned all EU travel to the US he wanted to be presidential. His unilateral ban without consultation undermined the stock markets. Donny from Queens ignorance of science is so apparent for everyone to witness.

8 million deaths from MDR Tuberculosis and 2 million deaths from Malaria annually put Covid-19 into perspective. SARS, MERS, MDR Tuberculosis and a recent Hanta Virus outbreak in China from rats reminds us that the world will continue without us.

Source: Coronavirus; que sait-on de L'épidémie? Le Monde Jeudi 27 Fevrier

Conclusion

Although microbiology is a relatively young science it has had an enormous impact on our health and well being. Without vaccines and antibiotics we would still be struggling to contend with epidemics of infectious disease and would be vulnerable to relatively minor infections. The microbes that make us sick are however vastly outnumbered by microbes that are essential for our existence and allow the Earth to remain habitable. For better or for worse, a world with microbes is unthinkable.

This book includes a series of articles concentrated on microbes lethal to mankind. Cancer features along with hepatitis C which may lead to cancer. Cancer cells with extraordinary growth behave in a malignant manner towards the human body and deserve a book fully dedicated to this killer disease. The Ebola outbreak and its eventual containment, an extraordinary disease, is caused by a fragile virus, easily killed by a 4% chlorine solution or even household bleach, devastated West Africa where victims died mostly from dire poverty and the lack of a health service. The massive western response significantly reduced the death rates in Liberia and Guinea, however Sierra Leone remains a cause for concern.

Covid-19 is a warning, the Earth may continue without humans, a more severe virus such as Marburg virus could finish mankind.

List of Illustrations

www.ingramcontent.com/pod-product-compliance
Lightning Source LLC
Chambersburg PA
CBHW072217170526
45158CB00002BA/630
9781326137960